A Brief History of Blood and Lymphatic Vessels

Andreas Bikfalvi

A Brief History of Blood and Lymphatic Vessels

Springer

Andreas Bikfalvi
Angiogenesis and Tumor Microenvironment Laboratory
University of Bordeaux and National Institute of Health
and Medical Research
Pessac, France

The work was first published in 2016 by EDP Sciences with the following title: Une brève histoire de vaisseau sanguin et lymphatique

ISBN 978-3-319-74375-2 ISBN 978-3-319-74376-9 (eBook)
https://doi.org/10.1007/978-3-319-74376-9

Library of Congress Control Number: 2018932419

Printed on acid-free paper

This Springer imprint is published by the registered company Springer International Publishing AG part of Springer Nature.
The registered company address is: Gewerbestrasse 11, 6330 Cham, Switzerland

Foreword

The vascular tree irrigates all tissues throughout the body, provides oxygenation and nutrient supply, and eliminates waste products. It is a system of considerable size and of paramount importance because, without a vascular connection, tissue dies. In humans, more than 80 diseases are linked to dysfunction of the vascular system. Understanding how the vascular tree is formed is therefore a major issue in medical research.

This book provides a historical view of important discoveries in angiogenesis research, information about current studies, and a discussion of still unresolved issues. It also contains a section of more conceptual and philosophical aspects related to the field. It is important that a recognized scientist actively working in this field write a synthesis on this topic.

In this book, Andreas Bikfalvi faced a triple challenge: To write a book that is interesting and readable for both the specialist and the general public, to describe original and new observations on the subject especially related to the history of different discoveries, and to discuss the more general aspects and conceptual issues of this research field. He has met this challenge admirably.

The book is indeed easy to read, is well illustrated, and deals with fundamental and translational aspects of vascular biology related to vascular development. It should therefore be of interest not only to biologists or medical students, but also to scientists active in this field as well as anyone interested in biology and medicine and in the history of science. It was a great pleasure to read it.

This book was published in 2016 in French, and it is now very timely that Springer Verlag publishes this English version to make it accessible to a wider audience.

Yale University
New Haven, CT, USA

Anne Eichmann

Preface

This book deals with blood and lymphatic vessels. Blood vessels are made of tubes of different calibers conducting the blood to the tissues and from tissues to the heart. Blood vessels are not inert ducts but are formed during embryonic development and endowed with plasticity during adult life. The formation of new vessels received the scientific term "angiogenesis" derived from ancient Greek angeon (αγγειον) and genesis (γενεσις).

If it was accepted that angiogenesis plays a role in embryonic development, its role in the development of different diseases was controversial. In this regard, the hostility of the scientific establishment toward the concept of tumor angiogenesis is revealing. In the 1980s, research on the cell cycle and oncogenes (tumor transformation-inducing genes) were dominant, and angiogenesis observed in tumors was considered a non-specific inflammatory reaction. At these times, any attempt at research on tumor angiogenesis was denigrated or even ridiculed, and it was difficult to obtain funding. Thanks to the tenacity of a man, Judah Folkman, considered the founding father of angiogenesis, the scientific establishment gradually changed its opinion and people hostile to the field became fervent defenders and even embraced the subject.

The idea of this book came to me in January 2008, when Judah Folkman was meant to attend a Keystone meeting in Denver, Colorado. We learned of his sudden death at Denver airport. This was a considerable shock for all researchers working in the field. For decades, Judah Folkman had given constant input and made angiogenesis emerge from the dark to become one of the flagship areas of biomedical research. I still remember very well the effects the reading of some of his work had on me. In particular, when I read one of his articles in the prestigious journal Science in 1984 on the effect of angiogenesis inhibitors, I was soon convinced that I must do angiogenesis research. The call was so strong that I abandoned my residency in hematology-oncology that I was doing in Germany at that time to come to Paris to work on angiogenesis.

The path taken following the initial work of Judah Folkman is considerable. Key factors and inhibitors of angiogenesis have been identified together with their mechanisms of action. Different mechanisms of angiogenesis have been defined and the cellular interactions that are involved in this process have been clarified. Inhibition of angiogenesis has been clearly validated clinically in cancer and in ocular and inflammatory diseases, and angiogenesis inhibitors have emerged in clinical use.

In Europe, a community of angiogenesis researchers has emerged, competing with laboratories across the Atlantic. Very fruitful collaborations have been established across Europe and conferences such as those we have organized with the European School of Hematology for over 10 years across Europe testify to this.

In France, we have set up a French Network of Angiogenesis, which has since 2008 became a society (the French Angiogenesis Society) and which brings together the main laboratories working in this field in France.

I would like to dedicate this book to my father Andras Bikfalvi Sr, a surgeon at the University of Giessen in Germany, who, apart from his profession as a thoracic surgeon, had a strong interest in research, especially in experimental surgery. He would certainly have liked to read my writings. I also dedicate the book to my children Alexis and Marianne, who are both physician and teacher. This book is a reflection of their respective interests. Finally, I would like to dedicate this book to our friend and colleague Jean Plouet, who died in 2010. Jean was a research director at CNRS and co-discovered with Napoleone Ferrara the growth factor of the vascular endothelium (VEGF, a major factor in angiogenesis) during his stay in the laboratory of Dr. Denis Gospodarowcz in San Francisco [1]. Jean was one of the co-founders of our French Angiogenesis Network. He had a very original and unconventional mind, which did not fit very well with the French academic system.

The story that is told in this book is closely related to my personal history with angiogenesis. Indeed, I have been a witness of and actor in most of the events described in this book alongside my cited colleagues, some of whom are my friends.

If I have not been able to cite all the significant works, it is not for any other reason than to keep a coherent narrative. My colleagues can forgive me surely. I wanted to avoid a too "academic" narrative and to remain comprehensible to a wide audience. I hope that this book, beyond the accessibility that I hope to offer, is of interest both to those who are versed in biology and medicine and to those who have no in-depth knowledge of the subject. For those who would like to go deeper into the subject, references that seemed to me essential are indicated at the end of the book.

This book is also, I think, a good introduction to vascular biology and angiogenesis for all those who want to start and pursue their research in these fields, and it is especially aimed at young students and post-doctoral scientists.

It should be noted that in this book I have focused primarily on blood vessel development and its role in physiology and pathology. I have not dealt with other important aspects, such as atherosclerosis and hypertension. I leave it to the experts in these fields to answer your questions.

This book was published in 2016 in French and my publisher (EDP Science) allowed me to proceed with an English version that is now published by Springer Verlag. The English version follows the structure of the French version, albeit several parts have been revised and updated. I published an article with regard to the conceptual analysis in the journal Angiogenesis (History and conceptual developments in vascular biology and angiogenesis research: a personal view.

Angiogenesis (2017) Nov 20(4):463–478) in 2017. Elements of this article have been incorporated in the last part (see Chapter 17 "Philosophy of the Vascular Tree") of this book.

I invite you to take an exciting journey through the history and the most current concepts concerning the research on the blood and lymphatic vessels. It probably did not escape you that the title of my book resembles the famous little book of Stephen Hawkins "A brief history of time". If I can touch you with my writings in the same way as this book by communicating my enthusiasm and awakening your curiosity, I have reached my goal.

Pessac, France
December 8, 2017

Andreas Bikfalvi

Contents

1 The Vascular System: What Is It?

Before going to the heart of the matter, the reader, if he or she does not already have a knowledge of biology, must familiarize him- or herself with a number of commonly used concepts and terminologies. The technical terminology and terms are easily understood and do not require specific knowledge. The reader may consult them at any time in the Glossary.

As already mentioned in the foreword, blood vessels are tubes of different calibers conducting the blood to the tissues and from the tissues to the heart (Fig. 1.1).

Blood vessels are not inert conduits; they are formed during embryonic development and have plasticity throughout adult life. Far from representing a uniform category, we must distinguish between the different types of blood vessels. The first is the arteries and veins, which have different structures and functions. The second subdivision of the vascular tree is a function of size and distinguishes between large vessels and smaller vessels called arterioles, venules, and capillaries (by decreasing size). Capillaries have an essential role in the exchange of nutrients and oxygen with the tissues. They are also heterogeneous and are distinguished as fenestrated and unfenestrated capillaries. Fenestrated capillaries have small openings that allow fluids and macromolecules to pass. They are mainly present in the kidneys and the lungs.

The structure of the blood vessels depends on their size; in other words, their form and function are different and constituted as microvessels or larger vessels. Capillaries consist mainly of endothelial cells, which may be covered with a layer of so-called mural cells or pericytes (Fig. 1.2). These cells are in contact with proteins forming a basal collagen membrane. The largest vessels are formed of several layers called intima, media, and adventitia.

The intima comprises the layer of endothelial cells and components of the extracellular matrices constituting the subendothelium. The media includes smooth muscle cells that have a contractile ability. The adventitia is made up of stromal cells and associated connective tissue. The stroma is the part of a tissue made up of matrix

A. Bikfalvi, *A Brief History of Blood and Lymphatic Vessels*,
https://doi.org/10.1007/978-3-319-74376-9_1

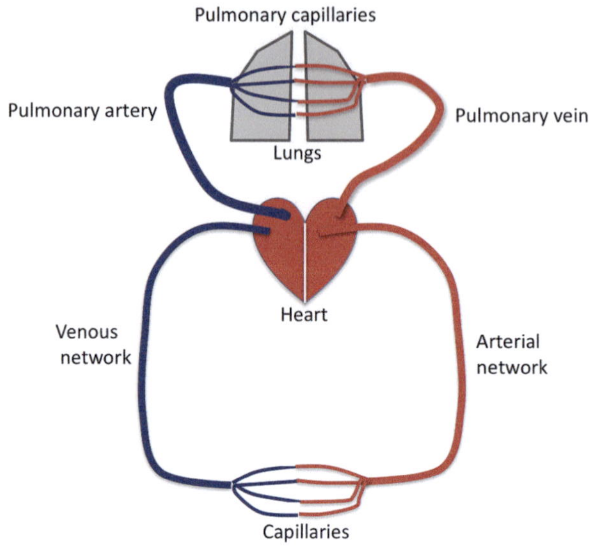

Fig. 1.1 The vertebrate blood circulatory system

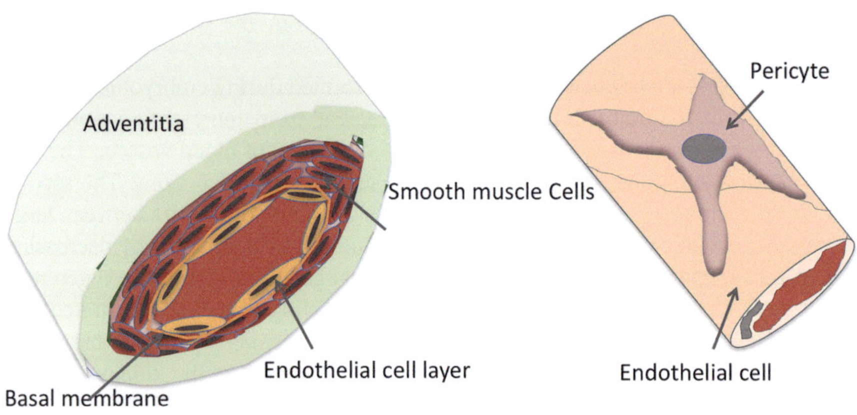

Fig. 1.2 Structure of large blood vessels and capillaries

proteins (collagen, fibronectin, etc.) and cells called fibroblasts. Fibroblasts are cells able to produce different matrix proteins.

Along with this closed circulatory system there coexists an "open" system (Fig. 1.3). The latter consists of lymphatic vessels that drain fluids from the tissues and play an important role in the immune defense of the body. Blood plasma is continuously filtered in the interstitial space, where the excess of liquid and macromolecules is drained through cell junctions made of endothelial cells. In addition, lymphatic vessels inside the intestinal villi absorb fat (lipids). Certain tissues such as skin and mucous membranes are exposed to foreign agents and antigens, and are particularly rich in lymphatic vessels.

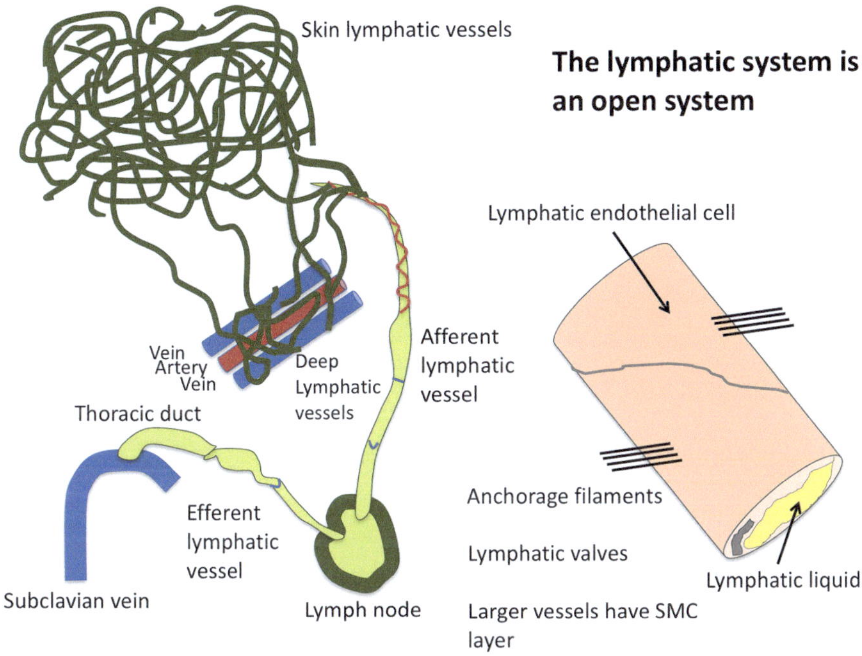

Fig. 1.3 Schematic representation of the lymphatic circulation and of the structure of a lymphatic vessel. Figure redrawn by the author, From Alitalo, Tammela and Petrova, Nature 2005, 438:946–953 [2]

Lymphatic capillaries are thin-walled vessels of about 30–80 μm diameter, composed of only one layer of lymphatic endothelial cells. Unlike the blood vessel capillaries, lymphatic capillaries generally do not have mural cells (pericytes). On the other hand, they have a discontinuous basal membrane which allows migration of immune cells such dendritic cells. Lymphatic endothelial cells also have discontinuous intercellular junctions, which allow the entry of lymphatic fluid and white blood cells into these vessels. Valves are another important structure found in lymphatics and allow the progression of the lymphatic fluid inside the lymphatic vessel, anchoring filaments made of collagen fibers and fixing the lymphatic cells to the matrix. In Fig. 1.3, the network of lymphatic vessels and their structure are shown.

2 History of the Vascular System

The vision of the vascular system has undergone many changes during history (Table 2.1). Galen, a native of Pergamon (131–201 BC), was the personal physician of Emperor Marcus Aurelius (Fig. 2.1). He described the circulatory system as a one-way system in which the veins and arteries are only distinguished by the quality of the blood they conveyed [3, 4]. He based his theory on those of predecessors including Erasistratus (304–250 BC), who postulated that veins conveyed blood and arteries only air. In this scheme, veins were connected to the right heart (atrium and right ventricle) and arteries, containing only air, to the left heart (atrium and left ventricle). Galen, in contrast, showed that arteries and veins both contained blood but of different qualities. For Galen, blood had its origin in the liver, passed from the right to the left heart to reach the major blood vessels to be transported to the various tissues, and finally was transformed into flesh. A fraction of the blood was also transported to the lungs, but only to nourish it, and another portion was transported from the right to the left heart through pores located in the septum. The air had the role of cooling blood and conveyed what was called "pneuma." In the heart, it was transformed into the vital spirit. The arteries conveyed the pneuma and veins and, in addition to blood, several humors (yellow and black bile).

Noxious vapors generated during the transformation of heat are ejected in a retrograde manner into the pulmonary veins during systole and then into the lung. Thus, the pulmonary veins are a ventilation device with a reversible function. The precise role of respiration and oxygen was not known at that time. The pneuma, which must revitalize the tissues, can be viewed as a sort of oxygen. We had to wait for Scheele, Priestley, and Lavoisier in the eighteenth century for the discovery of oxygen, which can be viewed as a molecular form of the vital spirit.

Leonardo da Vinci (1452–1519) had a special interest in the study of the heart and blood vessels and practiced anatomical dissection. However, he completely adhered to the Galenic principles of his time. As well as creating classical anatomical descriptions, Leonardo was conducting comparative dissections. During one of his dissections, he compared the blood vessels of an elderly man with those of a child [5, 6]. To quote Leonardo: "*The arteries and veins in the elderly that extend between*

A. Bikfalvi, *A Brief History of Blood and Lymphatic Vessels*,
https://doi.org/10.1007/978-3-319-74376-9_2

Table 2.1 Chronology of the discoveries of the vascular system until the nineteenth century

Name	Discovery
Galen (131–201)	Arteries and veins are separated
Leonardo da Vinci (1452–1519)	Comparative description of vessels of a young and old individual, hemodynamic description of heart valves
Ibn al-Nafis (1210–1288), Servetus (1511–1553), Columbus (1516–1559)	Description of the small circulation
Harvey (1578–1657)	General description of blood circulation and experimental evidence
Malpigni (1628–1694), van Loewenhook (1632–1723)	Demonstration of arterial and venous connections by capillaries
Ruysch (1638–1731)	The vascular tree is ubiquitous
Schwann (1810–1882)	The capillaries are composed of cells
His (1831–1904)	The cell layer covering the inner surface of vessels is called the endothelium
Rouget (1873), Zimmermann (1886), Mayer (1902)	Discovery of pericytes and distinction of different pericyte types

Fig. 2.1 Galen, physician of Marcus Aurelius. Image copyright free of licence, Bibliothéque Nationale de France (http://www.biusante.parisdescartes.fr/histmed/image?CIPB0985)

the spleen and liver were so thick which limits the flow of blood from the mesenteric vessels. Furthermore, vessels, apart from the thickness of their wall, grow in length and take a tortuous shape, similar to a snake. Shortly before his death, the old man told me he had lived 100 years . . . and I did the anatomical dissection to find the cause of this gentle death . . . . The other dissection was that of a child of 2 years where I found just the opposite . . . ." It is interesting to note that Leonardo described atherosclerosis, which can disrupt blood flow, leading to blockage of the vessels by thrombosis. It is amazing that the old man lived for 100 years!

Leonardo also carried out hemodynamic studies particularly on the aortic valves, studies that have remained valid [5, 7]. Leonardo was the first to study the small valves (mitral and tricuspid valves) and he identified their role in the regulation of blood flow. He conducted a detailed study of hemodynamic movement of the aortic valves and established the role of the sinus of Valsalva and the closure of the aortic

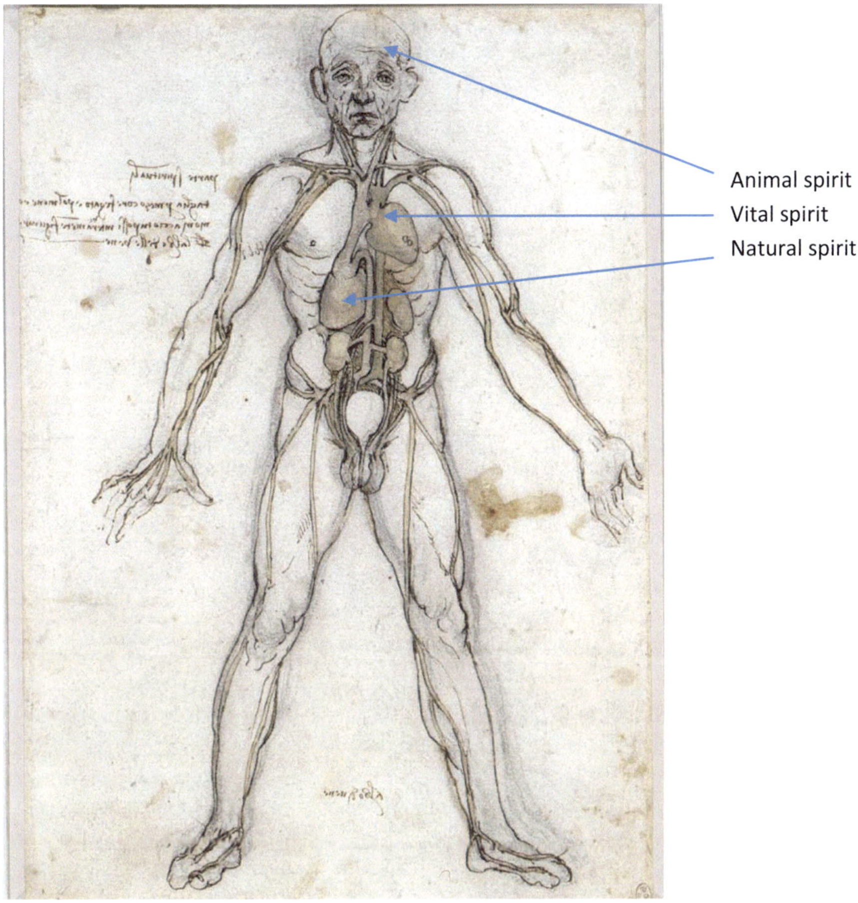

Fig. 2.2 The circulatory system according to Leonardo da Vinci. Theory of the three spirits. This drawing was made around 1490. The Galenic influence can be seen in this drawing (reproduced with permission: Royal Collection Trust/© Her Majesty Queen Elizabeth II 2017). The location of the three spirits is indicated in this figure. The natural spirit comes from the liver. It is then transported to the heart and peripheral vessels. The vital spirit is injected with the air and reaches the heart through the lungs. The animal spirit is generated in the brain. However, the brain tissue has no active role to play. The animal spirit is localized in the ventricles and cavities of the brain

valve. He also accurately analyzed vortex formation in propelling the blood through the valves. However, Leonardo did not revolutionize the understanding of how the vasculature and heart are organized or formed, and he adhered to Galenic medicine (Fig. 2.2). Technology was not sufficiently developed at that time to go any further, and the microscope had not yet been invented.

Arab medicine is claimed to have described the circulatory system much earlier [8, 9]. Indeed, Ibn al-Nafis (1210–1288) in "Sharah al Tashreeh al Qanoon" (Anatomical comments on the canon of Avicenna) wrote that the blood is

Fig. 2.3 Ibn al-Nafis, Michel Servet, Columbus (*from left to right*), discoverers of the small circulation. Ibn al-Nafis: from Takrouri MS and Khalaf M. Ibn al-Nafis contributions to science, Middle East J Anaesthesiol. 2003 Jun 17(2):163–176 [16]. GNU Free Documentation License. Servetus et Colombus. Images copyright free of licence, Bibliothéque Nationale de France. http://www.biusante.parisdescartes.fr/histmed/medica/page?08985&p=20. http://www.biusante.parisdescartes.fr/histmed/image?06015

transported from the left heart to the right heart through pulmonary circulation, and not directly through the pores lying at the cardiac septa (Fig. 2.3, left).

Michel Servet (Servetus) (1511–1553), who was burned alive by Jean Calvin in Geneva, had also formulated the idea of pulmonary circulation (Fig. 2.3, middle). In his book "Christianismi Resstitutio" [10, 11] he wrote: "*In this matter there must first be understood the substantial generation of the vital spirit which is composed of a very subtle blood nourished by the inspired air. The vital spirit has its origin in the left ventricle of the heart, and the lungs assist greatly in its generation. It is a rarefied spirit, elaborated by the force of heat, reddish-yellow (flavo) and of firey potency, so that it is a kind of clear vapor from very pure blood, containing in itself the substance of water, air and fire. It is generated in the lungs from a mixture of inspired air with elaborated, subtle blood, which the right ventricle of the heart communicates to the left. However, this communication is made not through the middle wall of the heart, as is commonly believed, but by a very ingenious arrangement the subtle blood is urged forward by a long course through the lungs; it is elaborated by the lungs, becomes reddish yellow and is poured from the pulmonary artery into the pulmonary vein. Then, in the pulmonary vein it is mixed with inspired air and through expiration it is cleansed of its sooty vapors. Thus, finally the whole mixture, suitably prepared for the production of the vital spirit, is drawn onward from the left ventricle of the heart by diastole. That the communication and elaboration are accomplished in this way through the lungs we are taught by the different conjunctions and the communication of the pulmonary artery with the pulmonary vein in the lungs. The notable size of the pulmonary artery confirms this; that is, it was not made of such sort or of such size, nor does it emit so great a force of pure blood from the heart itself into the lungs merely for their nourishment; nor would the heart be of such service to the lungs, since at an earlier stage, in the embryo, the lungs, as Galen teaches, are nourished from elsewhere because those little membranes or valvules of the heart are not opened until the time of birth. Therefore, that the blood is poured from the heart into the lungs at the very time of birth, and so copiously, is for another purpose. Likewise, not merely air, but air mixed with blood, is sent from the lungs to the heart through the pulmonary vein; therefore, the mixture occurs in the lungs. That reddish-yellow color is given to the spirituous blood by the lungs; it is not from the heart.*" Michael Servetus anticipated somehow the existence of a substance that gives color to the blood, which we know as oxygen. It should be noted that Servetus wrote this in a book of theology in which the person of Jesus and the meaning of the Trinity in the Christian religion was discussed.

Matteo Realdo Columbus (1516–1559) was a student of Andreas Vesalius and Professor of Anatomy and Surgery at the University of Padua and Pisa. In the second half of his life he became the personal physician of Pope Paul III. In his famous book "De Re Anatomica," Columbus corrected some errors that Vesalius had published in the *Fabrica* and also described how the blood does not go directly from the right heart to the left via pores located in the left ventricular wall, but through the lungs (Fig. 2.3, right) [12, 13]. He justified his presence in Rome with the Pope by saying that in Rome a large number of bodies was available to dissect, enabling him to examine and correct related theories, both ancient and modern (by his colleagues). Columbus was not very modest and probably had a high opinion of himself. Apart from the description of the cardiovascular system, he also wrote on the anatomical descriptions of female genitalia and the discovery of the clitoris!

There is a debate on the priority of the discovery of the pulmonary circulation between Servetus and Columbus [14]. It was insinuated that Servetus was a copyist. To quote Chéreau: "Servetus is only a copyist often unfaithful, sometimes awkward, always mystical" [15]. However, Servetus's manuscript had already circulated in 1547, long before the publication of "Re Anatomica," indicating the priority of Servetus over Columbus.

Andrea Cesalpinus (1519–1603) confirmed the existence of pulmonary circulation [17, 18]. It was argued by some that Cesalpinus was the discoverer of blood circulation because he would have already recognized the return of venous blood to the heart [19]. However, this seems to be incorrect. Cesalpinus thought that part of the venous blood was transported from the liver to the heart and then redistributed to the various tissues of the arteries. Another portion of the venous blood was directly channeled to the periphery to feed the various organs and tissues. In some ways, this places the center of the vascular system, in part, to the heart. This also postulates that, under certain circumstances, there may be a passage of venous blood to the arteries. However, this was not a major connection.

Fabricius Aquapendente (1537–1619), who was Professor of Anatomy at Pavia, is cited by most historians as having described for the first time the venous valves in 1579 [20]. However, in 1545 Charles Estienne had already mentioned venous valves when he described the "processes membranarum" in the veins of the liver [21]. Moreover, Lusitanus and Canano demonstrated their existence in public dissections in Ferrara in Italy and Jacques Sylvius described the valves in veins of the arms and legs in 1555. Their discovery was not considered at that time to be reliable, as Vesalius, the most famous anatomist of the time, could not find valves in veins, and the discovery was therefore attributed later to Fabricius. However, the role assigned to the valves was the opposite of what we now know. According to Fabricius and other anatomists, valves had the role of slowing down the centrifugal flow of the blood from the heart to the peripheral tissues. This was, they thought, because of gravity and of attraction of the blood by tissue. Valves were therefore a kind of safeguard device, which regulated the excessive movement of the blood distributed to the periphery. In 1627, Cesare Cremonini [22] postulated that the arterial blood not only conveyed the vital spirit and heath but also served to nourish the organs. Cremonini was a professor in Pavia and Averroen Aristotelian philosopher. Because of his pantheistic to atheistic beliefs, he was persecuted by the Inquisition. The Inquisition could not however punish him physically as Pavia was under Venetian jurisdiction.

All these were minor changes to Galenic theory until William Harvey. Even the discovery of the small circulation did nothing to modify the Galenic vision of the vascular system in this regard. It was William Harvey (1578–1657) (Table 2.1, Fig. 2.4) who rethought radically the role of the heart and blood vessels and who established the concept of the circulatory system [22, 23].

Harvey, who was also physician to King Charles I, based his theory on experimentation and not on reading and commenting on the theories of the ancients. Because anesthesia was something unknown at the time, Harvey practiced vivisection. This may shock the reader's sensitivity but the "Zeitgeist" or "mentality of the

Fig. 2.4 William Harvey around 1630. Image copyright free of licence, Bibliothéque Inter-Universitaire Santé (http://www.biusante.parisdescartes.fr/histmed/image?CIPB2093)

time," as one would say in English, was very different in the seventeenth century. The animal was considered to be devoid of soul and conscience and therefore much less sensitive to pain than men. In his treatise "Exercitatio Anatomica de Motu Cordis and Sanguinis in animalibus" [23, 24] dedicated to King Charles I (who was, by the way, beheaded during the Cromwellian revolution), Harvey showed that the vascular system was circulatory – arteries supplying blood to the tissues and veins carrying blood from the tissues. As early as 1616, Harvey had already formulated the concept of a circulatory system. Appointed in 1615, Harvey taught in the "Lumley and Caldwell" lecture series of the College of Physicians and gave there three lectures between April 16 and 18, 1616 [24]. In his second lecture, he uttered the following words: "*WH showed that the blood was continuously spread through the lungs to the aorta ...*," and then he said "*he demonstrates this using ligatures that blood passes from arteries to veins and shows therefore that blood movement is circular and heart beat are the cause of this phenomenon*" ("*WH observation per fabricam cordis sanguinem per pulmones in Aortam perpetuo transferri, finding per ligaturam transitum sanguinis ab ad arteriis venas unde A perpetuum sanguinis circulo motum in fieri pulsu cordis*").

Harvey performed his experiments on different animal species. Ligating the veins below the heart in two places of a fish, Harvey noted that the space between the ligature and the heart quickly becomes empty, indicating that the blood returns to the heart. Harvey found that when he seized the vena cava of a live snake between thumb and forefinger, the segment of the vena cava between the fingers and the heart was emptied of blood. The size and color revert to normal after lifting the obstacle. When, in contrast, the aorta is compressed, the heart swells and takes a deep purple or even pale color. These changes are reversible when the obstacle is removed. If the aorta of a dog or sheep is ligated to the base of the heart and the carotid or other artery is opened, the artery is found to be empty but veins are filled. This is consistent with the idea that arteries receive blood from veins only through the heart. He also conducted experiments by ligating the veins of the arms, irremediably leading to the interruption of flow (Fig. 2.5).

Harvey was also a visionary for the future of medicine and uttered in his reply to Jean Riolan, who criticized the theory of blood circulation, the following sentences:

"*The study of things which are normal in physiology is the first thing that doctors must learn. What is normal is the criterion for itself and for what is abnormal. By recognizing differences from this (normality) and its deviation from the natural, pathology becomes more apparent, and the practice and art of therapeutics which gives opportunity to discover multiple remedies*" [25]. This is clearly the foundation of modern medicine in which man is viewed in his globality. It is only the comparison with the normal condition for understanding the pathology and identifying therapeutic targets for which effective treatments can be developed. From this historical overview one could believe that the discovery of blood circulation is attributable to Harvey's predecessors, including Ibn al-Nafis, which has been claimed by some. However, Harvey differs from his predecessors in three ways: (1) he based his conclusions on observation and animal experiments, (2) his work had a much broader scope because it concerned both the small (pulmonary)

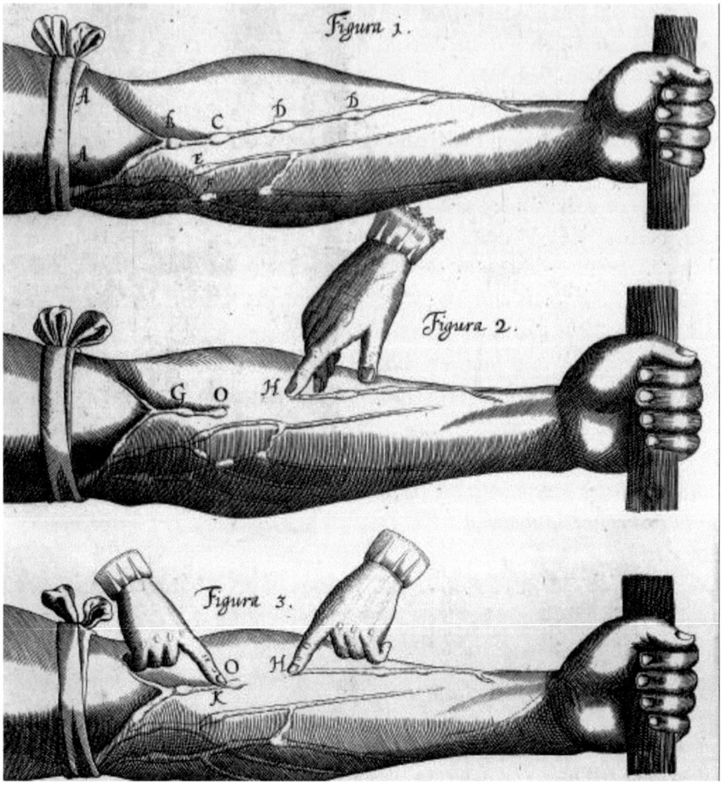

Fig. 2.5 Harvey's ligation experiment. The venous circulation is interrupted by a ligation, and then the index finger is placed on the trajectory of a vein. It is obvious that the blood does not circulate beyond the place where the index finger is placed. If a second finger is placed higher on the vein, it is noticed that the blood is not moving in the opposite direction. Harvey used a light ligation that only compresses veins and tight ligation that compresses veins and arteries. In the case of light ligation, venous blood reflux is only interrupted and the arterial pulse is still detected. In the case of tight ligature, there is no more venous reflux or forearm irrigation and the pulse is no longer perceived. Image copyright free of licence, Bibliothéque Inter-Universitaire Santé (http://www.biusante.parisdescartes.fr/histmed/image?01073)

circulation and the systemic circulation, and (3) it is important to note that neither Ibn al-Nafis, Servetus nor Columbus had separated themselves from the Galenic doctrine and had not assumed that circulation was a general anatomical and physiological fact, where veins were not distributing blood to the tissues but ensured the return of blood to the heart! According to them, only a fraction of blood was diverted to the pulmonary circulation to be routed to the left ventricle. The largest portion of the blood was distributed by the vena cava at the periphery. The lung was only an alternative route that allowed the passage of blood from the right heart to the left heart.

It should also be noted that Harvey's work had a general application for all living creatures with a vascular system and was not only limited to man as he reached these conclusions after studying a large number of animal species. However, it concerned only vertebrates in which a closed circulatory system exists in all species.

Thus, Harvey's work has clearly established the paradigm of the circulatory system. This is a major conceptual leap in the perception of the functioning of living organisms. The situation is similar to the problem of cell theory formulated by Schleiden and Schwann [26], as the theory was also preceded by previous observations. In general, for every scientific discovery, broad general concepts, which constitute a new paradigm in the classical sense, are often preceded by observations that illuminate a narrow aspect of the paradigm. These observations do not formulate a paradigm as far as a paradigm opens up a new conceptual framework for future research (see discussion about paradigms in the last part of the book).

What was this passage between arteries and veins in the tissue? William Harvey had cooked organs, including the liver, spleen, lungs, and kidneys, and had dissected tissues until he could see what he had called "capillamenta" or capillary filaments. These were in fact small arteries or veins because true capillaries are invisible to the naked eye. Harvey thought that either arteries and veins were connected through anastomosis (vascular connection between larger vessels) or blood was passing directly through the pores of the tissues.

This question was not specifically answered until Malpighi's (1628–1694) studies and the use of the microscope, which allowed the visualization of the structures we now call capillaries (Table 2.1, Fig. 2.6). His discovery of the capillaries was

Fig. 2.6 Marcello Malpighi, the discoverer of blood capillaries (Copyright Licence-free Image, Bibliothéque Inter-Universitaire Santé (http://www.biusante.parisdescartes.fr/histmed/image?CIPA0613) and an image of the edition of "De pulmonibus observationibus anatomicæ" of Thomas Bartholin from 1663 which include two letters from Malpighi to his friend Borelli. From: Bartholin, Thomas: De pulmonum substantia & Motu diatribæ. Accedunt Cl. V. Marcelli Malpighii de Pulmonibus observationibus anatomiacæ. Hafniæ [Copenhagen], typis Mathhiæ Godicchii, sumptibus Petri Hauboldi, 1663. Image courtesy of the Hagströmer Medico-Historical Library

published as letters to Borelli 'De Pulmonibus' and reprinted in 1661 in Leiden in the Netherlands. These letters were published by Thomas Bartholin in "De motu pulmonorum substantia & diatribæ" in 1663 [27]. Malpighi wrote in these letters: "*By this means I could clearly see that the blood flows through tortuous vessels and the blood is not spilled in tissue spaces but still pushed through tubules and distributed at the level of multiple branches.*"

Malpighi was from Bologna and was an enlightened spirit who managed to combine clinical activity with research. Indeed, he continued to see patients at the most intense moments of his research career. Because of his activity, his character, and the history of his family, he generated a lot of hostility. In 1684, his house was burnt down in Bologna, destroying his microscopes and many of his manuscripts. These attacks were repeated twice, fortunately sparing Malpighi who was not at home at the time of the attacks. These criminal acts were attributed to a Sbaralea, Professor of Anatomy at the University of Bologna, for both personal (the brother of Malpighi had killed the brother of the Sbaralea) and intellectual reasons because, for the Sbaralea, the detailed anatomical observation of plants and animals did not contribute anything to the evolution of medicine (see "from recentiorum Medicorum Fluidiis" of Sbaralea printed in Göttingen) [28]. The microscope was for the Sbaralea an inadequate tool for studying physiology and pathology.

Malpighi's observations were confirmed in an independent manner by Antonie van Leeuwenhoek (1632–1723) (Table 2.1, Fig. 2.7). Antonie van Leeuwenhoek also used a microscope of his own fabrication made by simple lenses held in a metal device. He studied the vascularization of different species including tadpoles, fish, rooster crests, rabbit ears, and bat wings. He wrote: "*If the passage of arterial blood to the veins of the tadpole is now clearly demonstrated not to have occurred in other structures than in vessels, we can infer that this is also true for the human body and other animals*" [29].

Using injections with wax, the great Dutch anatomist Frederik Ruysch (1638–1731) showed the presence of blood vessels in almost all tissues of the body (Table 2.1, Fig. 2.8). He also described the vasa vasorum and bronchial

Fig. 2.7 Portrait of Antoni van Leeuwenhoek (1632–1723). Image copyright free of licence. Bibliothéque Inter-Universitaire Santé (http://www.biusante.parisdescartes.fr/histmed/image?02473)

Fig. 2.8 The anatomical lesson of Frederic Ruysch. Image copyright free of licence, Bibliothéque Inter-Universitaire Santé (http://www.biusante.parisdescartes.fr/histmed/medica/page?111502x1913x12&p=271)

capillaries. In 1696, Ruysch proposed that the tissues are nothing more than a set of different prodigious structured vascular networks [30]. He attributed the difference in the function of tissues to the different structures of the vascular networks in various organs. Although vascularization plays an important role in the function of tissues, this view is obviously incorrect, as the specific function of a tissue is controlled by the different functional cells that compose it. Frederik Ruysch also described for the first time the presence of lymphatic valves. He also had one of the most famous European collections of anatomical specimens whose different parts were kept in a special "home-made" liquid of which he had the secret formula. This collection was sold to Frederick the Great for 30,000 Guilds during his lifetime, a considerable sum then.

However, it was not known then, and not until the early nineteenth century, how the capillaries were structured, whether they had a cell wall, whether they were only made of fibrous materials, or whether they were just simple channels inside tissues without a fibrous membrane.

To illustrate these misconceptions, a few examples are given. In 1835, J. W. Earle described the capillary membrane without channels as being similar to "*brooks in the moist earth*" [31]. This was also supported by the observations of Wedemeyer who noted that "*In long arteries, blood winds down in the membrane without channels, channels being formed of the substance of the tissues. The blood in the finest capillaries no longer flows in real vessels whose walls are not formed by a membranous substance, which differs from the adjacent tissue in texture and compactness, but in single channels, the walls of which are formed by surrounding cells*". In 1820, Ignaz Döllinger (1770–1841) (not to be confused with the German theologian Ignaz von Döllinger), Professor of Anatomy and Embryology at the University of Würzburg, said that the blood was surrounded by mucus, the basic

Fig. 2.9 Theodore Schwann, father of the cell theory. Image copyright free of licence, Bibliothéque Inter-Universitaire Santé (http://www.biusante.parisdescartes.fr/histmed/medica/page?08985&p=154)

substance in tissues. He stated that "*the blood flows like water in a field and is not enclosed in a tube*" [32, 33]. In 1831, Marshall Hall described the capillary network in many organs and showed that their diameter was constant [34]. He also wrote that capillaries are "simple channels" as opposed to real tubes. In 1833, Johannes Müller stated in the first edition of his famous book "Elements of Physiology" that the capillary wall was made of single thickened edges and did not have a membrane [35].

The capillary wall was observed for the first time in the tadpole by Theodor Schwann (Table 2.1, Figs. 2.9 and 2.10). In 1839 and 1847, Schwann was the first to describe what was later to be called the endothelium [26]. He wrote that the capillary vessels in the tail of young or fully-grown tadpoles were surrounded by a thin membrane, clearly visible, which showed no fiber arrangement. The thickness of this "membrane" was not uniform in various capillaries and not clearly visible. Most importantly, Schwann observed cell nuclei at different locations of the capillary wall that were, for him, either nuclei of cells that strictly belong to the capillaries or, alternately, nuclei from adjacent epithelial cells. Schwann was more inclined to the first explanation. Schwann in his own words wrote: "*Well individualized nuclei at different places and this in young fully developed tadpoles. These nuclei appear to lie within the thickness of the wall or at the inner face of the wall. This allows two types of explanation. Either these nuclei are nuclei of capillary cells, or they belong to epithelial cells that invade the capillaries. These nuclei often appear to be at the inside of the vessel and should be much more abundant if they originated from epithelial cells. Thus, it seems likely that these nuclei are derived primarily from cells of capillaries, although the various arguments do not definitively decide the issue.*" If we follow Schwann in his arguments, he finds the nuclei in two different places of the capillary wall, inside the wall and near the lumen of the capillary vessel. He has correctly analyzed that the cells were proper cells of the capillary wall. However, he did not exclude a participation of epithelial cells.

Fig. 2.10 Schwann's drawings from 1848 representing, among others, a blood capillary. The two cell types, internal and external, which were identified by Schwann at this time, are recognized. The internal cells correspond to what are now called endothelial cells and the outer cells to what are called pericytes (image reproduced from Schwann, 1848 [26], copyright free)

In the original drawing of Schwann, shown in Fig. 2.10, we see that one nucleus of a cell he said was in the capillary wall is instead found in more external position. Another cell nucleus is rather located in an internal position of the vessel wall. In light of our current knowledge, the cell located in the external position is likely to be a pericyte, the inner cell in contact with the vascular lumen certainly being an endothelial cell. Although Schwann made the mistake of referring to the possibility of the presence of epithelial cells in the wall, he assumed correctly that all cells belong to the vessel wall. However, he was not able at this time to distinguish the two cell types of the capillary wall, endothelial cells and pericytes.

The internal layer of cells was later called "endothelium" by His (1831–1904) [36]. The distinction into two types of capillary cells (blood and lymphatic capillaries) was made in the early eighteenth century by Albert von Koelliker (1817–1905) [37]. Koelliker was Professor at the University of Würzburg and a convinced anti-Darwinist. His contributions to developmental biology and to nervous system were important and he was particularly in favor of the neuronal theory of organization of the nervous system. His book "History of the development of man and the higher animals" ("Entwicklungsgeschichte des Menschen und der höheren Tiere") was very influential at the time.

The cell theory of the vessel wall of the capillary was not accepted immediately. However, Schwann had supporters. Even Johannes Müller, who was hostile at the beginning to this idea, changed his mind. He wrote in the third and fourth edition of his famous book "Elements of Physiology" [38, 39]: "*Schwann recently demonstrated by the use of the microscope that the capillaries were not only covered by membrane tissues, but have their owned structures in which round entities like the nuclei of epithelial cells are present*" [38]) and then in a later edition he stated more clearly that capillaries are not only channels but actually have a membranous wall composed of cells ("*capillary vessels are not mere channels in the substance, but they possess membranous walls composed of cells*").

Nevertheless, the controversy continued. In 1849, Bennett Dowler, a physician from New Orleans, exclaimed about one of his colleagues: "*Professor Carpenter's anatomical history of the capillaries does not seem embarrassed with any doubts whatever, though it bears on its face very little that can be called absolute certainty.*" He says that *"the capillary circulation is carried on through tubes which have distinct membraneous parieties originating in cells!"* In another place, he says that *"the capillaries arise from a minutely anastomosing network, into which the blood is brought by the ramifications of the arteries on one side and from which it is returned by the radicles of the veins on the other." Cells!, network!, ramifications!, radicles!, one side!, and the other side!* [40]. Apparently, new theories were struggling to get accepted across the Atlantic at that time! The disproving of previous concepts or theories by new evidence often leads to resistance, and it takes time for evidence to become generally accepted, in particular when it supports a paradigm shift, and this is what cell theory represented for scientists at that time.

As mentioned, capillaries are not only made of endothelial cells but also of pericytes. These contractile structures at the level of capillaries were described for the first time by Charles Marie Benjamin Rouget in 1873 in the capillaries of the hyaloid of the frog and other tissues [41]. The hyaloid membrane of the frog is a richly vascularized membrane of the eye. He described structures arranged longitudinally made of irregular cell bodies surrounding capillary vessels. Although he could not color these cells with vital dyes, he postulated that they were muscle cells. The clear identification of these cells was made by Zimmermann in 1886 and Mayer in 1902 [42, 43]. Zimmermann invented the name "pericytes" for the cells and used silver nitrate for staining them, whereas Meyer favored methylene blue for visualization. Zimmermann distinguished three types of pericytes [42]: precapillary pericytes, capillary pericytes, and postcapillary pericytes. Pre- and postcapillary pericytes have morphological transition [42] to smooth muscle cells. These latter pericytes were classified as transitional pericytes to distinguish them from real capillary pericytes. Precapillary pericytes have a number of extensions that tend to wrap around the vessel. Capillary pericytes are elongated spindle-shaped cells extending in a longitudinal manner on the vessel and have many short secondary processes. Postcapillary pericytes have a smaller star shape. Given that Zimmermann had no modern imaging tools, he could not perform three-dimensional visualization of the capillaries, as can be done today. He would have seen, as is the case for the capillaries of the retinal vessels, that capillaries can also be completely surrounded by pericytes.

Zimmermann had also postulated that the contraction of pericytes could control vascular permeability. The role of pericyte in permeability is now confirmed, but it is rather the degree of contact of pericytes with the vascular tube that influences vessel permeability. This is the case for tumor vessel or for diabetic retinal vessels, as we see later in this book.

Table 2.1 summarizes the different stages of discoveries of the circulatory system. These discoveries are spread over a period of almost 2,000 years! There was a latency period with some advances until the sixteenth and seventeenth centuries, followed by an acceleration in discoveries up to the nineteenth century. Harvey, in this evolution, is the link between the medieval and modern thinking with the introduction of experimental methods. Since the nineteenth century, the acceleration of discoveries has become exponential because of technological development. However, we must remember that the Harveyan revolution was not a result of a technological revolution but of a fresh look at how science should be done by promoting empiricism and the experimental method over philosophy and the authority of the ancients.

Thus, to sum-up what we have discussed here, we can see that the idea of the organization and function of vessels has dramatically changed over history and has undergone multiple changes. These changes are primarily the result of the introduction of the experimental method and the refusal to believe blindly in ancient theories and explanations. It is to emphasize that life and medical sciences were very dependent on religion and theology as the example of Michael Servetus shows. To summarize this part of the history of the discoveries of the vascular system, we can identify four phases – a Galenic phase, a Harveyan phase, a capillary phase, and a cellular phase. Figure 2.11 provides a representation of the evolution of these concepts through history.

We must now turn to another aspect of the circulatory system, namely the explanations of the role of blood vessels in nutrition and tissue inflammation. In the first part of the nineteenth century, two theories existed to explain nutrition of tissues and inflammation (Fig. 2.12). The first theory was called the "blastemal theory." According to the blastemal theory, the blastema, a fluid that, among other

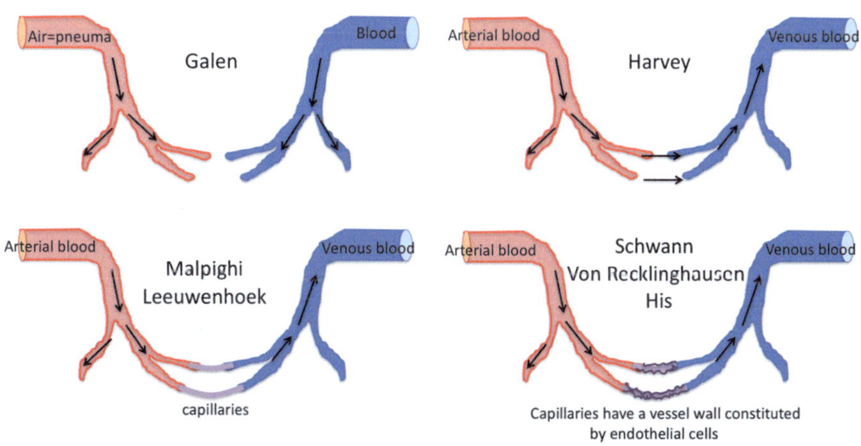

Fig. 2.11 Schematic representation of early discoveries of the vascular system. Redrawn by the author from [44]

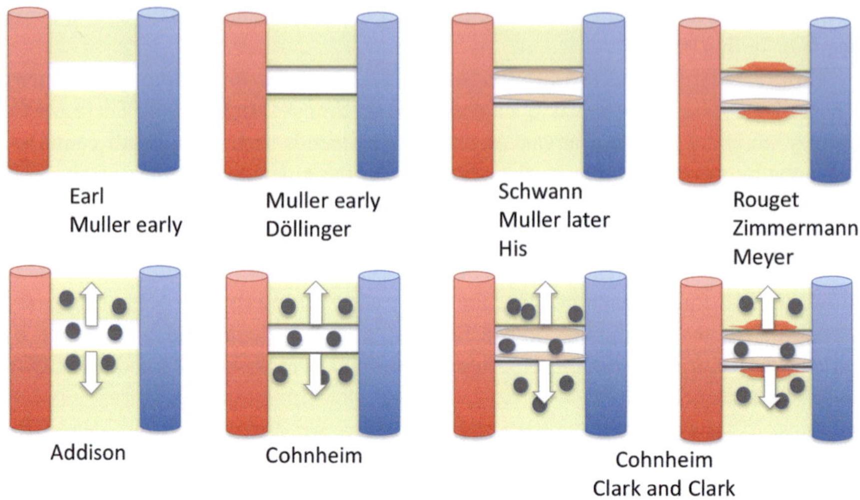

Fig. 2.12 Evolution of theories and concepts concerning the interaction of the blood with the vascular wall. At first, it was not believed that a real capillary vessel wall exists. This vision changed gradually and led to the vision we have today of the capillary structure which includes endothelial cells, a basement membrane, and a pericyte layer. At the same time, the corpuscular concept was introduced which opposed the concept of the blastema. In the corpuscular theory, blood cells are transported from the blood to tissues to give rise to cells and pus (for white blood cells). In the initial version of this theory, it has been asserted that the existence of a vascular wall is incompatible with the transmigration of cells from the blood. This concept was gradually transformed and led to the vision we have today, which gives the endothelium the leading role in this process. Figure by the author

things, contains fibrin and albumin, is transferred from the blood to the tissues to feed them and to contribute to the formation of normal and inflammatory cells (pus cells). The other theory, the "corpuscular theory," postulated that circulating cells (red cells, white cells and blood cells) continuously circulate between the blood and the extravascular space, contributing to the formation of cells in inflamed tissues under physiological and pathological conditions.

The physiologist and English physician John Hughes Bennett (1812–1875) and defender of the blastemal theory, described the capillary wall as composed of a firm, continuous, and transparent membrane, without leaving any opening, with the presence of nuclei at different locations [45]. This membrane retains particles (solid bodies), not allowing their passage through the capillary wall. This fluid, after its passage into tissues, contributes to tissue or pus cell formation.

One of the main supporters of the corpuscular theory was the English physician William Addison (1802–1881) [46, 47]. Addison denied the existence of a clear membrane of the vascular wall. He stated "*The capillaries are channels that are surrounded by living tissue islands. Remove these islands you will see that none of these vessels persists*". He postulated that the "so-called" vascular wall was made of fibers derived from white blood cells that then merge with those tissues. According to Addison, in normal conditions the particles adhere to the fibrillar border of the vascular channels and either disintegrate into fibers or are incorporated into the "vascular wall". They become, after their passage, structural tissue components or secretory cells. Inflammation may be associated with more accumulation of leukocytes at the border of the capillaries, ultimately resulting in the formation of pus. However, Addison's theory was speculative and not supported by experimental data. The corpuscular theory implied for him the lack of a real vessel wall, otherwise the cells would be unable to transmigrate into the tissues. Those who postulated the existence of a real vessel wall were, themselves, opposed to the corpuscular theory, which requires the passage of cells through the vascular wall. Finally, experimental evidence that transmigration was possible in an intact vessel wall, was provided, at least for leukocytes, by Julius Cohnheim (1839–1884), who showed that white blood cells pass through the vascular wall and are transformed into pus in tissues [48, 49].

Today we know that the supporters of the two theories were right and wrong! The scientist couple Eliot and Eleanor Clark, who we meet later in this book, were the first in the 1900s to show that the endothelium plays an essential role in the adhesion and transmigration of leukocytes across the vessel wall [50]. It is interesting to note that Addison's theory contains a modified "metaphoric" version of the old Galenic theory that the blood becomes flesh once arrived in tissues. This is obviously false in the strict sense. However, there are conditions where this may be observed. During cancer metastasis, tumor cells transmigrate through tumor vessels and circulate in the bloodstream to reach other tissues. Having arrived at their destination, they transmigrate again into tissues to form secondary tumors, which represent pathological tissues.

It should be noted that all these discoveries describe blood vessels as static entities and say nothing about their development. Yet we know that the embryo develops and the child grows up and, thus, organs and tissues must grow consequently. This must, of course, be accompanied by a concomitant vessel growth. Although developmental biology had progressed in the nineteenth century and twentieth, in particular thanks to the German school of developmental biology, these pioneers did not have enough knowledge to formulate the concept of biology as a "chemistry of life." Therefore they did not know that factors stimulating the growth of blood vessels could exist.

The first description of in vivo blood vessel growth was made by the Scottish surgeon John Hunter (1728–1793) (Table 2.1, Fig. 2.13). Hunter was a quite remarkable character. He was not only a surgeon and anatomist but also conducted many kinds of research, ranging from comparative anatomy to the pathophysiology of venereal disease [51]. He is credited with having made the largest number of dissections in the eighteenth century. He observed that blood vessels develop after ligation which describes for the first time the phenomenon of angiogenesis.

John Hunter had conducted a classic experiment in a stag. In July 1785, Hunter ligated one of the carotids arteries which irrigated one side of the reindeer antler. The deer wood concerned became cold, indicating that it did not receive enough blood. Hunter expected the wood, which was thus devoid of blood, to die. However, after 1–2 weeks, the wood became warm again. He thought that the ligature had failed, which was not the case. On the contrary, he saw circulation provided by new vessels

Fig. 2.13 Portrait of John Hunter (*left*). Popliteal aneurism of a coachman prepared and injected by Hunter (*right*). This was the first case that Hunter had treated with his method of femoral artery ligation. This was a 45-year-old coachman who, 6 months after surgery, found the full use of his leg. However, 1 year later this patient died of a disease that was not related to his popliteal aneurism. Reproduced with permission from the London Museums and Royal College of Surgeons (© Museums at the Royal College of Surgeons)

(which is now called collateral circulation) which irrigated the wood. Hunter had the opportunity to apply this finding immediately to patients. At that time, coachmen wore very tall boots that rubbed at the popliteal cavity and induced popliteal aneurysms. At that time, amputation of the affected limb was performed, in this case, of course without anesthesia, which led to a likely death. Hunter thought that if the blood circulation of the femoral artery was interrupted at a certain level, the aneurysm would no longer be fed by the blood and would regress [51–53]. Moreover, he postulated that a collateral circulation could form which compensated for the absence of flow through the femoral artery. This circulation would be sufficient to irrigate the rest of the lower limb. This was indeed the case, and Hunter saved the leg of a significant number of coachmen who would otherwise have been doomed to certain death. Hunter therefore showed for the first time that vessels had a plasticity and could be newly formed when specific constraints were imposed on them.

However, Hunter had no idea how this was happening mechanistically. It is possibly that Hunter speculated on possible causes, but the science of the time did not allow him to test further these ideas experimentally.

John Hunter is not only famous for his research in angiogenesis – he is also well-known for his work in anatomy, biology, surgery, and medicine in general. He contributed to comparative anatomy and determined, among other things, that tumors could be removed by surgery before invading adjacent tissues.

In 1849, the Scottish physician John Hughes Bennett, of whom we have already heard (see above), said [54]: "*The Blastema necessary for this purpose is first obtained from the original exudation poured out, but after a time when the fibrous tissue increases, new vessels are formed in it, which continue to furnish materials to the new growth in the same manner as the old vessels furnished materials to the old tissue. Notwithstanding, the property inherent in the exudation itself, formally alluded to, pathologists are obliged to acknowledge that a tissue once formed and furnished with blood vessels, possesses the properties of exerting a kind of attraction on the blood, whereby such matters are transuded through the capillaries...* ." As can be seen, Bennett states that vessels are formed in the tissues and that they nourish the tissue, but he does not say that vessels are attracted by an active process. He says that the tissue attracts blood, and because Bennett was an adept of the blastema theory, it was the non-corpuscular transudate which was attracted.

Carl Thiersch (1822–1895), Professor of Surgery who, at that time, worked in Erlangen in Germany, described the formation of new vessels in the tumor stroma (Fig. 2.14). He discovered that new vessels originate from pre-existing vessels. Thiersch wrote in a paragraph his luminous intuition about the importance of an interaction between tumor epithelial cells and vessels [55]: "*The epithelium depends on the vascular stroma in the same way as a soil in which it has taken root. Like the plant, the epithelium brings its autonomous potential for development and growth and only demands to supply those substances which are necessary for its development*" ("*Das Epithel hängt eben von der gefässhaltigen Stroma nicht anders ab, wie die Pflanze vom Boden in der sie wurzelt. Wie die Pflanze, so bringt auch das Epithel sein selbständiges Entwicklungs und Wachstumsvermögen mit sich und verlangt wie diese weiter nichts als die Darbietung jener Stoffe deren es zur Entfaltung seiner Gestaltungs und Absonderungsfähigkeit bedarf*"). Everything is said in this paragraph! Thiersch has already anticipated this close relationship between tumor and vessel. Nevertheless, he had not, in his days, formulated the hypothesis that active demand of tumor cells for vessels necessarily involves the existence of a (biochemical) mediator produced by the tumor cells.

Thiersch was not only famous for his work on the vascular system but also showed that metastases occur by dissemination of tumor cells in the body. Further observations for physiological angiogenesis were subsequently made, such as those of the pathologist Arthur Tremain Hertig (1904–1990), the father of pediatric pathology, who, in 1935, described placental angiogenesis in female monkeys in gestation [56].

Eliot R. Clark, following Remak, Sigmund Mayer, and others, first proposed mistakenly that the endothelium was contractile (the contractile part of the vessel is made up of smooth muscle cells). He has, however, correctly observed the development of vessels by cell division from the cells of the vascular wall [57]. Clark was

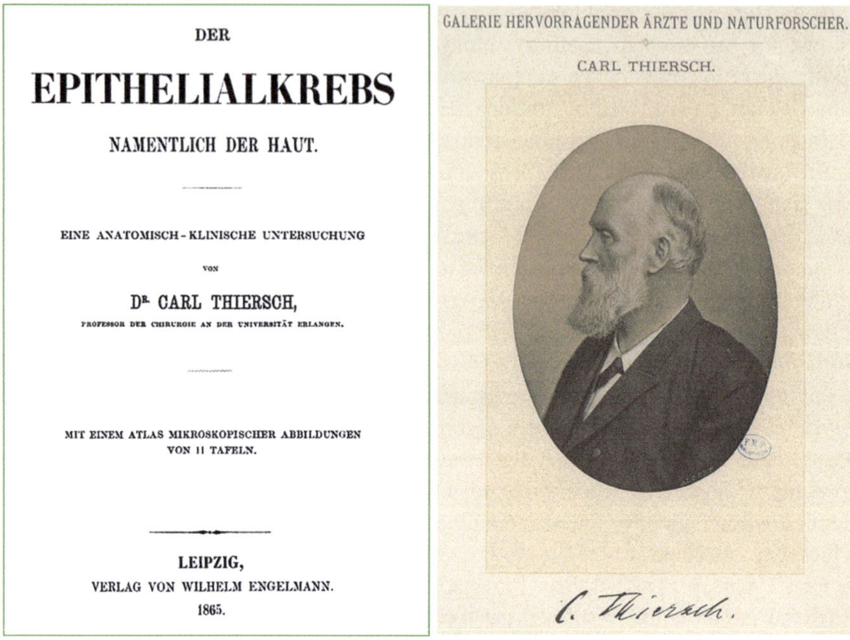

DER

EPITHELIALKREBS

NAMENTLICH DER HAUT.

EINE ANATOMISCH-KLINISCHE UNTERSUCHUNG

VON

DR. CARL THIERSCH,

PROFESSOR DER CHIRURGIE AN DER UNIVERSITÄT ERLANGEN.

MIT EINEM ATLAS MIKROSKOPISCHER ABBILDUNGEN
VON 11 TAFELN.

LEIPZIG,
VERLAG VON WILHELM ENGELMANN.
1865.

Fig. 2.14 Carl Thiersch and the first description of tumor angiogenesis; (*left*) permission to reproduce image, free copyright; (*right*) licencing-free image, Bibliothéque Inter-Universitaire Santé (http://www.biusante.parisdescartes.fr/histmed/image?CIPB0715)

able to observe vessels over several days and found that the vascular wall (more precisely the cells of the vascular wall) emitted tiny protoplasmic processes, which he called buds or sprouts. He postulated that cell nuclei migrate after division into these buds and then develop into new vessels. If this was the first observation of the phenomenon of vascular budding or sprouting, the interpretation of this observation is partially correct, as we see later (the sprouts are in fact made up of cells that emit filipodia and nuclei do not move into the filipodia).

Paul Broca (1824–1880) (Fig. 2.15) who, apart from his work on the nervous system, was also interested in cancer, had already observed that vessels were detected in tumors, but did not attribute to them an active role in tumor development [58, 59]. Only cancer cells had an active role to play by breaking into the veins and lymphatics, thus explaining metastatic dissemination. His description mainly concerned the invasion of the largest vessels, leaving the capillaries aside. Broca was a convinced supporter of the analysis of pathological processes using the most advances technologies of the time. He wrote in his thesis in 1849 that: "*any observation on cancer not subject to observation by the microscope must be considered to be null and void.*" The name of Paul Broca is also strongly associated with neurology and anthropology, disciplines of which he was a relentless promoter. Paul Broca was, besides being a scientist, a free thinker and a committed humanist.

Fig. 2.15 Paul Broca who also contributed, besides neurology, to different aspects of oncology. Image copyright free of licence; Bibliothéque Inter-Universitaire Santé (http://www.biusante.parisdescartes.fr/histmed/image?anmpx35x0017bis)

He was agnostic or atheist by conviction and a remarkable mind which our times would greatly appreciate.

Ernst Goldmann had made fundamental observations on the angiogenic process in tumors ("tumor angiogenesis") (Table 2.2) in two articles published in the "The Lancet" in 1907 and in the Proceedings of the Royal Society of Medicine [60, 61]. Goldman goes further than Thiersch and his predecessors, and clearly describes the phenomenon of sprouting of capillary vessels. He bases this on observations in humans and animals using, in particular, X-ray imaging. He wrote more precisely: "*The normal blood vessels of the organs in which the tumor is developing are disturbed by chaotic growth, there is a dilation and spiraling of the affected vessels, marked capillary budding and new vessel formation, particularly at the advancing border.*" Goldmann carried out these experiments in man and in animals. To view the capillary network newly formed, he injected a mixture of bismuth oil into the femoral artery and took X-ray images (Fig. 2.16). These observations clearly show the presence of vascular proliferation in tumors. It is evident that the technique used by Goldmann did not allow the visualization of capillaries in a precise manner and, in fact, he observed larger vessels in the tumor. However, it is clear from the images that tumors attract vessels and promote their ingrowth.

Goldmann's interpretation of these observations was curious at the time (his interpretation was only recently validated for a certain type of blood vessels called high endothelial veinules; see later in this book). He saw in angiogenesis a mechanism of defense against tumor growth and not an enhancing factor or a promoter of tumor development. To quote Goldmann: "*I regard vascular neoformation as a standard by which we may test the body's power of reacting against malignant tumours.*" *Therefore I regard the network of newly formed blood-vessels merely as*

Table 2.2 Early historical developments in tumor angiogenesis

Name	Discovery
Thiersch, 1869	First description of emerging tumor vasculature from pre-existing vessels. Formulation of the "soil and seed" hypothesis
Goldman, 1907	Description of tumor angiogenesis and formation of vascular buds ("sprouts") in tumor vessels
Clark and Clark, 1939	Demonstration of morphological characteristics of blood vessels in the transparent chamber of the rabbit ear
Ide, 1939	Demonstration of tumor angiogenesis in epithelioma tumor model in the transparent chamber of the rabbit ear
Greene, 1941	Observation of vessel growth by transplanting rabbit tumors to animals of other species and use of the anterior chamber of the eye as an inoculation site
Algire and Chalkey, 1945	Observation of differences in the effect on neoangiogenesis of tumor tissue compared to normal tissue using the transparent skin chamber in the cat
Mervin and Algire, 1956	Observation of the difference in the effect on neoangiogenesis of a tumor tissue compared to a normal tissue implanted in the muscle. Demonstration of the capacity of tumor tissue to induce an angiogenic response
Greenblatt and Shubik, 1968	Observation of neoangiogenesis in the transparent chamber model in the hamster equipped with a filter. Demonstration of the existence of a soluble factor
Ehrmann and Knoth, 1968	Observation of neoangiogenesis using the chicken embryo model (use of filters). Demonstration of the existence of a soluble factor
Folkman, 1971	Definition of the paradigm. Postulate of the existence of diffusible factors regulating tumor angiogenesis

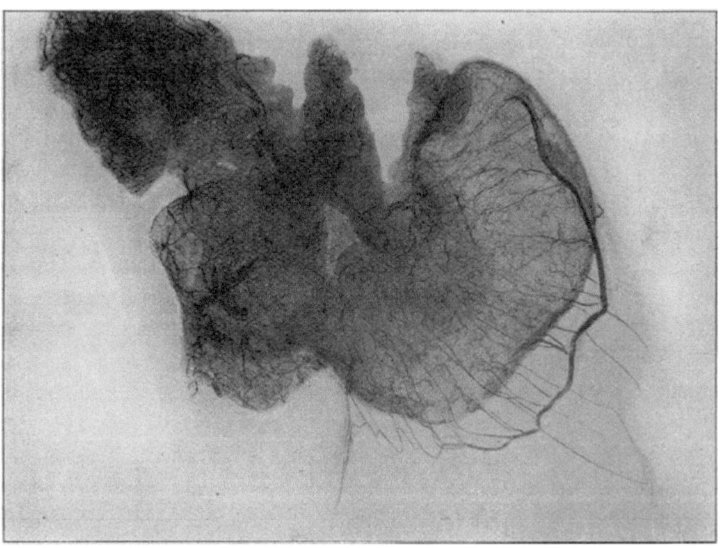

Fig. 2.16 X-Ray of the stomach of a patient with stomach cancer injected with bismuth oil. There are multiple new vessels migrating to the tumor. Permission to reproduce (Sage publication licencing)

useful in producing more active blood circulation. Intensified circulation itself, if I may so call it, is the effect of all those healing powers which I have just referred to, inclusive of inflammatory agents, such as those recently recommended for the treatment of cancer by Bier and others. The efficacy of this intensified circulation is naturally dependent upon the presence of defensive factors in the blood. This was revised by Judah Folkman as we see later.

I would like to mention some curious observations I made at the reading of the original Goldmann articles. The two articles published in The Lancet and the Proceedings of the Royal Academy of Medicine are strictly identical. This is no longer allowed according to present day standards of publication, but this was surely different at the time. Furthermore, in many of the secondary literature, the reference of Goldmann is badly quoted and his name anglicized. However, Goldmann was a Professor at the University of Freiburg in Germany, even though he published his works in English scientific journals and his name was still spelled with two ns (Goldmann). These errors are introduced and propagated in the literature when only secondary sources are consulted and not the original articles.

Goldmann's observations were reinforced by Greene in 1939–1941 (Table 2.2) [62, 63]. He wrote: "*A series of experiments was undertaken in an attempt to transplant rabbit tumors to animals of alien species using the anterior chamber of the eye as an inoculation site. The growth characteristics of the transplanted tumors were generally similar to those observed in the natural host, but noteworthy exceptions occurred. The tumors obtained a blood supply from the foreign host and invaded the periorbital tissues*" ... Then he noted [64]: "*The growth rate increased following vascularization, and at the present time the transplants are approximately five times their original size.*"

In 1939, Clark and Clark (Table 2.2) [65] improved the experimental technique by using a device known as the "transparent chamber of the rabbit ear." They were able to study more precisely the characteristics and morphological changes of blood vessels. In 1945, Algire and Chalkley (Table 2.2) [66] used a transparent skin chamber implanted into the skin of a cat to study angiogenesis secondary to injury or implantation of normal or tumor tissue. They showed that the angiogenic response induced by the tumor was more important and earlier than that induced by normal tissue or after an injury. They concluded that the growth of a tumor is closely linked to the development of a vascular network. In addition, Algire and Lagallais reported later observations on the growth and vascularization of melanomas [67]. In 1956, Merwin and Algire (Table 2.2) [68] found that growth of normal and neoplastic tissues transplanted into the muscle was not significantly different in the beginning, but a significant difference was observed longer term. Normal tissues induced a limited vaso-proliferative response, whereas tumor cells induced the formation of new vessels outgrowing the implant. The intensity of the response seemed to be influenced by the distance between the implant and the host tissue. Tumor implants had a much greater capacity because they could induce new vessels at a distance greater than 50 μm, which normal tissues could not do. In 1939, Ide assumed that soluble factors produced by tumor cells are responsible for the growth of blood vessels [69]. The experimental evidence was only later provided. Indeed, Greenblatt and Shubik [70] had implanted a so-called “Millipore” chamber (porous membrane) in a hamster and placed some tumor fragments inside it. In this type of experiments, cells are not able to cross the chamber and only liquids can pass through. They observed that the tumor mass rapidly invaded the entire chamber. New vessels had also formed. This could only be because of diffusible factors (diffusing through the membrane) released by the tumor. These results were confirmed by Ehrmann and Knoth [71] who used the chick Chorioallantoic membrane (CAM) embryo model. This membrane is richly vascularized and is used by the embryo for gas and waste exchange. After fixation of the tumor fragments on millipore filters (which do not let tumor cells pass) and implantation of these membranes on the CAM, the formation of new vessels was observed (for more details on the CAM, see later in the book).

These observations had thus paved the way for works whose aim was to characterize the factor(s) involved. This is described in more detail in the next chapter. The idea that inhibition of vascularization could be used as a paradigm for cancer treatment was not yet conceived, and the characterization of angiogenesis factors not yet undertaken. However, I have found in the writings of Paul Broca a therapeutic approach that aims to coagulate blood in vessels and thus to destroy the tumor [59]. This method, called galvanopuncture, consisted of destroying the tumor by implanting electrodes and applying an electric current. This leads to an obliteration of the blood flow and destruction of vessels and tumor cells. In Paul Broca words: “... *This posed, we used the galvanopuncture to fulfill two distinct aims: 1) to make coagulates in vascular tumors, aneurysms, varicose veins, erectile tumors; 2) to disrupt the tissue of solid tumors*”... and he continues “... *I have cured by this means, unwittingly, a tumor as small as an almond, which occupied almost the whole extent of the upper lip of a child still breast feeding. My regrettable colleague Jamain had sent me this child to the hospital, where I replaced Mr. Nelaton. The tumor made rapid progress. I decided to treat her by galvanopuncture, hoping to obtain only the coagulation of the blood. I plunged into the tumor four needles, three of which were placed in communication with the positive pole of a Volta battery; the fourth needle was connected alone with the negative pole. The discs were only forty-three millimeters in diameter, and, with such a low intensity, I thought to be able to carry tension up to fifty pairs. It was too much. Before the coagulation of the blood was complete, I perceived that an eschar had occurred around the point of the negative needle, at a very small distance from the skin. I stopped at once. The scarification appeared at first rather thin, but the very next day I could make sure that it had four millimeters in diameter. When it detached, it left a cavity wide and deep; and the rest of the tumor changed so well, that the lip of the child gradually returned to its normal volume.*”

It should be noted that Broca had not made an explicit link between coagulation in the blood vessels and the deleterious effect on the tumor cells themselves. One hundred and thirty years later, Philip Thorpe (University of Texas, Dallas) used a molecular approach to target tumor vessels in order to coagulate the blood [72]. This approach, called vascular targeting, aimed at destroying tumor vascularization, is discussed later in this book.

Taken as a whole, these observations clearly provided a description of tumor vasculature, and suggest that the interaction between the tumor and vascularization is important for the development of cancers.

Table 2.3 History of the discoveries of the lymphatic system

Name	Discovery
Anselli (1581–1626)	Observation of lacteals in the intestine of the dog
Pecquet (1622–1674)	Description of the main lymphatic vessels and the thoracic duct
Rudbeck (1630–1702)	General description of lymphatic vessels
Bartholin (1616–1680)	Invention of the term "lymphatic"
William (1718–1783), John Hunter (1728–1793)	The lymphatic system is the only "absorbent" system
Monro secundus (1733–1817)	The lymphatic system is "absorbent" and distinct from blood vessels
Hewson (1739–1774)	Detailed description of the lymphatic system and circulating elements in the lymphatic vessels
Ludwig (1816–1895)	Study of the composition of the lymphatic fluid
Von Recklinghausen (1833–1910)	Identification of lymphatic endothelial cells
Sabin (1871–1953)	Centrifugal "venous" theory of lymphatic vessel formation from veins
Clark, Clark, and Clark (1933–1937)	Observation of the growth of lymphatic vessels and budding (sprouting)
Oliver (1999)	Identification of the "master" regulator prox1 for differentiation of the lymphatic vessels
Alitalo (1996)	Discovery of lymphangiogenic factors and their receptors (VEGF-C, VEGFR3)
Alitalo (2001), Detmar (2001)	Evidence that lymphatic neoangiogenesis has a role in the dissemination of lymphatic metastases and involvement of VEGF-C in this process

Apart from the "classical" vascular system, there is another system called the lymphatic system, as mentioned above (see Table 2.3 for a summary of the history of discoveries). Lymphatic vessels were observed as early as 300 BC. by Herophilos and Erasistratus of the Alexandrian school. However, the existence of the lymphatic vessels was actually demonstrated for the first time in 1622 by the Italian Gasparo Aselli (1581–1626), a professor at the University of Pavia in intestinal preparations from the mesentery of the dog (Fig. 2.17) [73]. He observed in the dog, after food intake, lacteales which actually represent small vessels, and which contain a whitish creamy liquid. He believed, however, that the lymphatic fluid was drained into the liver according to the galenic medicine of his time.

The main lymphatic vessels, the thoracic duct, and their connections with the blood circulation were described for the first time by Jean Pecquet in 1651 [74]. A coherent and more detailed description of the lymphatic system was given by Thomas Bartholin [75] and Olaf Rudbeck [76] in 1653 (Fig. 2.18). There was a lively dispute between Bartholin and Rudbeck about this discovery, for Rudbeck had already presented his results in a public dissection in the presence of Queen Christine of Sweden in 1652. However, if it seems that Rudbeck had preceded Bartholin, the term "lymphatic" was invented by Thomas Bartholin in 1653 (Table 2.3).

Fig. 2.17 Gasparo Aselli, discoverer of lymphatic vessels. Permission to reproduce (Bibliothèque Nationale de France). Image of the book cover courtesy of the Hagströmer Medico-Historical Library

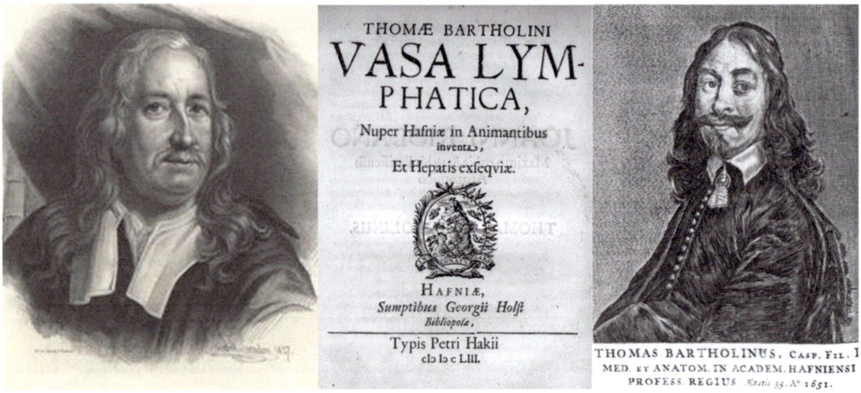

Fig. 2.18 Olaf Rudbeck (*left*). Cover of Bartholin's Vasa Lymphatica (eds. Copenhague: G. Holst, 1653) (*middle*). Thomas Bartholin (*right*). Permission to reproduce: Uppsala universitetsbibliotek, Carolina Rediviva Kart- & bildenheten/Maps & Pictures. Middle and right images copyright free of licence, Bibliothéque Inter-Universitaire Santé (http://www.biusante.parisdescartes.fr/histmed/image?02046; http://www.biusante.parisdescartes.fr/histmed/image?00157)

We met with John Hunter earlier. It is remarkable that this man not only contributed to better knowledge of blood vascularization but also made important observations concerning the lymphatic vessels. In 1758 he was the first to discover the lymphatic vessels in the necks of birds and formulated with his brother William the theory that these vessels were merely absorbing [77]. Later, William Hunter, with his assistant William Cruikshank, postulated that the lymphatic system was the only absorbing system and carried out a series of experiments in the presence of his brother and other "gentlemen scientists." This work was pursued by William Hewson (1739–1774), a student of Hunter, who described the lymphatic vessels of reptiles and fishes. He also observed the presence of solid ("cellular") elements for the first time in the lymphatic vessels. William Hewson gave an excellent definition of the lymphatic system. He said: "*The lymphatic vessels are small pelucid tubes which have now been discovered in most parts of the human body; the fluid they contain is generally as colourless as water, a circumstance which procured them first the name Ductus Aquasi, and afterwards that of Vasa lymphatica. The course of the lymph, like that of the chyle, is from the extreme parts of the body towards the center, and the lymphatic vessels commonly lie close to the large blood vessels*" [78].

To understand the composition of the lymphatic fluid, Karl Ludwig (1816–1895) developed a technique for collecting the lymphatic liquid from different parts of the body [79]. He thought that the lymphatic fluid was a filtrate derived from blood, which is partially correct. Carl Ludwig headed the Leipzig Institute of Physiology, which was at that time one of the most advanced institutes in biomedical research. He is the father of modern cardiovascular physiology and the source of many findings, especially in the cardiovascular field [80]. Similar to blood vessels, lymphatic vessels are also made of endothelial cells, which was demonstrated by von Recklinghausen in 1862 [81]. He had made these observations using silver nitrate staining that showed that the lymphatic capillaries were bordered by a thin layer of polygonal endothelial cells.

It was commonly accepted that lymphatic vessels had their origin in the tissues and then converged toward the veins. This view was overthrown by the observations of Florence Sabin (Fig. 2.19, left) [82, 83]. By an elegant series of experiments, she showed, in contrast, that the lymphatic vessels are derived from the veins (Fig. 2.19, right). In mammals during embryogenesis there is the appearance of structures called sacs and cisterna. Two pairs of sacs are found at jugular (near the neck) and iliac (near the hips) locations, two non-even pairs of sacs are found in the retroperitoneal (in the dorsal position) position and at the "Cisterna Chyli." The lymphatic vessels then bud/sprout from these sacs. To quote Florence Sabin: "*At first the lymphatics were thought to begin in wide cavities in the walls of the various cavities of the body, and then, as these openings proved difficult to find, attention became focused on the relation of the lymphatics to the tissues. The theory which I hold is that the lymphatics arise from the endothelium of the veins and grow by multiplication of endothelial cells.*" Using pig embryos injected with a specific dye at different stages of their development, Sabin showed that the lymphatic vessels emerged by budding from veins. She also showed that the lymphatic system was a unidirectional system collecting lymphatic liquid from the tissue that was then drained to the veins.

Fig. 2.19 Florence Sabin (*left*). Permission to reproduce Vermont Academy J.&C. Salzman Library. Network of lymphatic vessels in the skin of a pig embryo (*right*). This is achieved by injecting dyes into the pig embryo to trace the origin and development of the lymphatic vessels. The injections of the dyes are made at different embryonic stages to allow visualization of the centrifugal aspect of the expansion of the lymphatic vessels. Redesigned by the author and modified from Fig. 214 of Sabin (1902) [83]

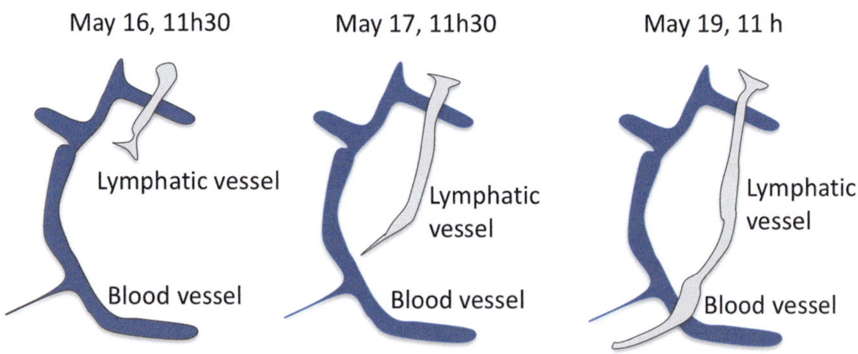

Fig. 2.20 Representation of Clark's drawing of lymphatic vessels and formation of lymphatic buds/sprouts in the frog larva. The drawing represents the successive stages of bud/sprout formation (according to Clark ER: Anat Rec, 1909, 6:261 [84], redrawn and modified by the author)

These results were reinforced by additional observations of lymphatic vessels development in the tail of the frog larva or by using the transparent chamber in the rabbit ear (Fig. 2.20). Observing the growth of lymphatics in the tail of the frog larva, Clark (1909) found that fine projections of the lymphatic vessel wall were visible and that the tip of the emerging capillary terminates at a pointed end [84]. The tip of the capillary also changed permanently. These observations were reinforced in mammals by studies using the transparent chamber of the rabbit ear [85–87]. Clark and Clark also observed for the first time the formation of lymphatic buds ("Sprout") which then grew to form a tubular structure with a lumen inside.

It should be noted that Florence Sabin was the first women member of the American Academy of Sciences.

However, the exclusive venous origin lymph vessel has recently been questioned, although the latter remains predominant. Indeed, it has been shown that a fraction of lymphatic vessels may be formed at peripheral locations (see Chap. 13 "Lymphangiogenesis Enters the Dance"). Recent discoveries of lymphangiogenesis factors and receptors are described later in this book.

Lymphatic vessels are invaded by tumor cells. As Paul Broca describes in his treatise on tumors [59]: "... *Another time, examining under the microscope a small axillary ganglion, large only as a pea, and engorged as a result of breast cancer, I saw in this ganglion a number of cylindrical casings, more than a tenth of a millimeter wide, and distended by a large quantity of cancer cells. This observation, which I have communicated to the Anatomical Society, and which I have since repeated on several occasions, succeeds only when ganglions engorged for a very short time; the specific elements of the cancer soon destroy the walls of the lymphatic vessels and spread uniformly throughout the ganglion*." However, in this description, the lymphatic vessels have a purely passive role, exposed to the onslaught of tumor cells. Until very recently this concept was generally accepted by the scientific community. This vision is now largely changed and it is now recognized that lymphatics play an active role in metastatic dissemination.

Evolution of the Vascular System

3

Eukaryotic organisms are divided according to the usual taxonomy of protozoa and metazoans. Metazoans can be classified either as vertebrates and invertebrates or as diploblasts and triploblasts. Vertebrates have an internal cartilage framework and bone, which invertebrates lack. The diploblast/triploblast distinction is made according to the presence of two or three embryonic leaflets from which tissues and organs form. Diploblasts do not possess more than two leaflets (endoderm and ectoderm), whereas triploblasts possess three (endoderm, ectoderm, and mesoderm).

Any unicellular or multicellular organism must feed, oxygenate, and eliminate carbon dioxide and metabolism waste products. In the simplest organisms, this occurs by simple diffusion. However, diffusion is effective only over relatively small distances. Any increase in size is accompanied by an increase in surface, but this is not proportional because the size (or rather the volume: cubic radius, r^3) of the organism increases faster than the surface (proportional to the square radius, r^2). At one point during evolution, the surface does not constitute an effective diffusion area and therefore imposes a selective pressure on the organism. This leads to an adaptive response, either by increasing the surface by folding and compaction accompanied by an adaptation of the metabolism of the organism, or by internalizing the exchange surface and the constitution of a circulatory system. Simple multicellular organisms such as platyhelminthes feed and oxygenate themselves by simple diffusion. However, when the organism becomes more complex, this is no longer sufficient, and an internal circulatory system develops. The internal circulatory system appeared in the ancestor of triploblasts more than 600 million years ago as a way to overcome time constraints (speed at which nutrients, etc., reach cells) and diffusion distance. Triploblasts possess, in general, two circulatory systems – a so-called coelomic circulation and a vascular circulation. The coelom is located between the outer wall and the intestine; it is filled with liquid and lined with mesothelial cells. The

The original version of this chapter was revised. An erratum to this chapter can be found at https://doi.org/10.1007/978-3-319-74376-9_19.

A. Bikfalvi, *A Brief History of Blood and Lymphatic Vessels*,
https://doi.org/10.1007/978-3-319-74376-9_3

coelomic fluid is transported by the contractility of mesothelial cells (called myoepithelial cells). In some species, the coelom can be subdivided into different compartments by partitioning. this not only allows gas and nutrient exchanges and waste treatment, but gives sufficient space for organ movement and increase of organ size. Moreover, the coelomic space represents a buffer zone (cushion) to external physical constraints and has a role in posture. In mammals, the coelomic space exists only in a vestigial form and thus no longer ensures the exchange functions observed in less-evolved species.

The vascular system, on the other hand, is filled with blood and is intended to provide a guided distribution of the latter to the tissues. In invertebrates, vascular tubes do not have an endothelial cell layer but only possess a matrix that is in direct contact with the blood. The movement of the blood is ensured either by contraction of the myoepithelial cells or by muscularized contractile pumps (the heart). The vascular system is either closed or open. In invertebrates, both types can be observed. Open systems are observed in insects, arthropods, and some molluscs. The blood, which circulates (hemolymph), is injected into the cavity (hemocoel) of the animal. In the case of the open system, tissues located more distantly from the heart theoretically receive less oxygen or nutrients than the more proximal tissues because the blood pressure drops quickly after being pumped through the heart. In insects, the distribution of oxygen is assured not by the circulatory system but by a finely branched tracheal system. In this case, the circulatory system is used primarily for nutrient distribution and for the recovery of waste. Closed systems are present in some invertebrates and in all vertebrates, except for the lymphatic system which is open and connected to the blood-vascular system (see Chap. 13 "Lymphangiogensis Enters the Dance"). The basic principle of closed systems is that exchange occurs in a specific location (capillaries in tissues) and this in a uniform manner. The distribution is no longer a function of the distance of the tissue or organ from the heart. In invertebrates, vascularized spaces are located between the coelom and the ectoderm on one side, and between the compartimentalized coelomic cavities on the other.

The inner lining (luminal surface) of the vessel consists of a basal membrane produced by epithelial (ectodermal) or mesothelial cells(coelomic), which is different in vertebrates where the inner surface is covered by an endothelium (Fig. 3.1). In

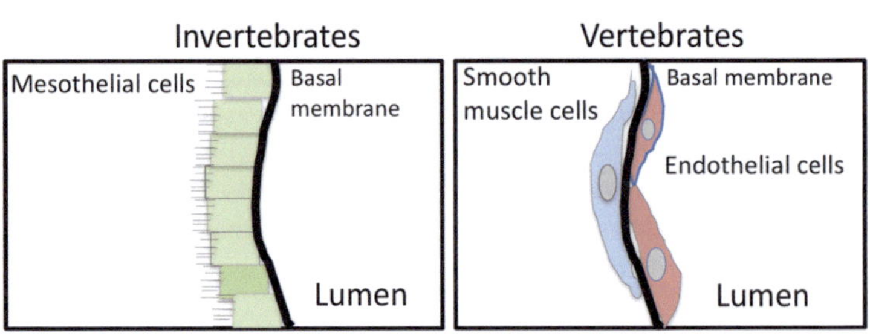

Fig. 3.1 Comparative structure of a vessel of vertebrates and invertebrates (redesigned by the author and modified from Monahan-Earley, R. Dvorak, A.M. Aird, W.J. Thromb Haemost 11 Suppl 1, 46–66 [88]). It should be noted that the invertebrate vessel does not have an endothelial cell layer. The vascular tube is, in this case, composed of epithelial cells and the matrix is exposed to the luminal surface of the epithelial tube

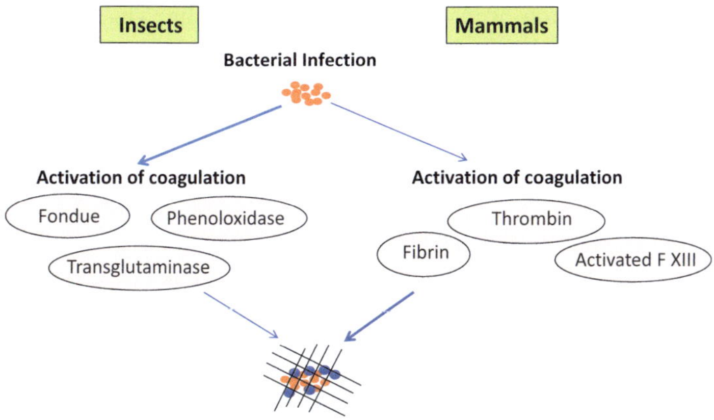

Fig. 3.2 Blood coagulation in vertebrates and invertebrates (*Drosophila*). Drawing by the author modified from Theopold U, Krautz R, Dushay MS (2014) Dev Comp Immunol 42:42–46 [91]

invertebrates, the blood is propelled either by the contraction of the mesothelial cells (myoepithelial cells) or by the cardiac pump, which in the case of invertebrates has only one chamber. For further details, the reader may consult the excellent article by William Aird, head of the vascular biology center at Harvard University [88].

As already mentioned, it should be noted that the extracellular matrix and the basal membrane are in direct contact with the blood in invertebrates. In vertebrates and mammals this would lead to adhesion of blood platelets and activation of coagulation. Thus, one may wonder why coagulation is not activated in invertebrates.

To understand this phenomenon better, we need to discuss briefly the coagulation system of vertebrates and invertebrates [89–91] (Fig. 3.2). Blood coagulation in vertebrates occurs after adhesion and platelet activation through a cascade of biochemical reactions that involves numerous pro-enzymes, enzymes, and inhibitors. This results in the formation of a stabilized crosslinked fibrin structure because of the involvement of enzymes called transglutaminases. Fibrin is then resorbed through the intervention of the fibrinolytic system. In blood coagulation there are two activation pathways, an intrinsic and a so-called extrinsic pathway. Both pathways converge in the activation of prothrombin to thrombin, which finally cleaves the fibrinogen to convert it to fibrin. The intrinsic pathway is initiated by an activation system involving a factor called factor XII and kallikrein. The extrinsic pathway involves a factor called tissue factor. There are many regulatory loops involved in coagulation, such as anti-thrombin III or protein C. Fibrinolysis is regulated by cascade activation of a number of molecules including plasminogen tissue activator (TPA) and plasminogen urokinase activator. These enzymes convert plasminogen into plasmin, the enzyme that ultimately degrades fibrin in order to resorb the clot.

In invertebrates, the activation systems are somewhat different. There are no proteins homologous to vertebrates but other molecular actors are involved. Three species of invertebrates are representative, with different molecular regulations.

In the crab, a cascade is initiated by the Lipopolysaccharide (LPS), which induces the activation of a protein called protein C. This is followed by a reaction cascade leading to the conversion of the coagulogen into coagulin. This coagulin is then crosslinked by a transglutaminase. Protein C is present in the hemocyte membrane. Another activation is caused by the interaction of β-glucan with a protein called G protein. This activation pathway converges with that initiated by LPS at a pro-complex, which then initiates the transformation of coagulogen into coagulin. Thus, the mechanism is similar to that of vertebrates but it uses different actors, the pro-complex playing the role of prothrombin/thrombin in invertebrates.

In crustaceans, the system is simpler. There is no proteolytic cascade, but only the crosslinking of a plasma dimeric protein (of a still unknown nature) by two transglutaminases released by hemocytes.

In *Drosophila*, some candidate molecules involved in coagulation have been identified [91]. One of these proteins is called "molten" protein, a transglutaminase substrate that regulates, as we have seen, the crosslinking of the "molten" protein. Other proteins are Eig71Ee (gp150) and hemolectin. How these proteins interact to form a coagulum is still poorly understood.

Invertebrates produce a number of molecules with fibrinolytic properties. However, in most cases these molecules have been tested for their property of lysing blood clots from vertebrate blood such as silkworm cocoonasis or earthworm lumbrokinase. Little is known about the lysis of coagulin or the "molten" protein.

It may be thought that hemocytes perform some function similar to blood platelets by regulating coagulation. Indeed, in *Drosophila*, two populations of hemocytes called plasmatocytes and crystal cells are activated rapidly after bleeding and release prophenoloxidase in the form of crystals. This enzyme produces tyrosine derivatives of small molecular mass that participate outside of the transglutaminase to the crosslinking of coagulogen. Transglutaminase allows loose crosslinking whereas the mechanism that involves prophenoloxidase allows the coagulum to harden [91]. As we have seen, in vertebrates, adhesion of platelets to the subendothelium, which is formed by the basal membrane and matrix molecules, activates blood platelets and initiates the thrombotic process. One may wonder why hemocytes do not adhere to the matrix that is exposed naked throughout the invertebrate vasculature. It is possible that anti-adhesive factors are present in the hemocoel as well as strong anticoagulating factors.

The endothelium evolved in the ancestral vertebrate animal about 500–600 million years ago to optimize blood flow and to serve as a barrier. This probably occurred after evolutionary divergence between urochordae (such as the squid) and cephalochordae (such as the amphioxi). Moreover, the endothelium constitutes a surface, which makes it possible to localize immune and coagulation functions. As already mentioned, the endothelium is present in vertebrates and absent in invertebrates, and invertebrates have only an extracellular matrix that serves as a barrier. It should be noted that in the older concepts of the capillary structure (see Chap. 2 "History of the Vascular System"), vessels were devoid of endothelial cells, which is reminiscent of an invertebrate vasculature. Some invertebrates, such as cephalopodia, have cells on the luminal surface, but these cells do not have tight junctions and are not firmly attached to the capillary membrane.

What is the origin of endothelial cells from an evolutionary point of view? Endothelial cells can theoretically originate from the differentiation of circulating cells (hemocytes or amoebocytes, as seen above in amphioxus) or mesothelial cells (myoepithelial cells). Hemocytes/amoebocytes express precursors of the Platelet-derived Growth Factor, (also called PDGF) and VEGF (Vascular Endothelial Growth Factor) receptors [92, 93]. It has been shown, for example, that in *Drosophila*, migration of hemocytes is dependent on VEGF and PDGF [92, 93]. Three VEGF homologues are expressed there and are located along the hemocyte migration pathway. The three forms of VEGF and their receptors are essential for the migration of hemocytes because their mutation abrogates it. This suggests that the ancestral function of VEGF is to control cell migration. This function has been co-opted by endothelial cells because it is essential to the formation of the vascular tube. The other possibility of the origin of the endothelial cells is trans-differentiation of mesothelial cells. As seen later, the epithelial vascular tubes express a homologue of the VEGF receptor. The debate on the evolutionary origin of endothelial cells is therefore not definitively solved.

Branching morphogenesis is a process encountered in invertebrates and vertebrates. Angiogenesis of vertebrates has incorporated the general molecular mechanisms of this morphogenic process. It is remarkable that homologues of VEGFs or FGFs (fibroblast growth factors) have been identified in some invertebrate species such as insects or annelids. However, the mechanisms that regulate the expression of VEGF and its tissue localization appear to be different. As already mentioned, VEGF receptors are not expressed in invertebrates in endothelial cells (because the endothelium is absent) but are present in hemocytes and, in some cases, in the myoepithelial cells.

In this context, the comparison with *Drosophila* (vinegar fly) is particularly revealing (Table 3.1). *Drosophila* has a hemocoel and an open circulation represented by the vascular tube and a tracheal system. The tracheal system resembles, with regard to its structure and morphological and molecular changes during development, the vertebrate vascular system. Air enters by diffusion through the tracheal system, which ramifies finely and delivers oxygen to the organs. The development of the tracheal system depends on oxygen and uses molecular pathways similar to angiogenesis to induce budding and tracheal ramification. In both cases, a "guide" cell (tip cell) is present that responds to a gradient of stimulating factors. However, the nature of the stimulatory factor is different. In *Drosophila*, the factor is represented by Branchless/FGF. In vertebrates, it is the VEGF system that takes over this function. As in vertebrates, Branchless/FGF is

Table 3.1 Comparison of regulators of sprouting in vertebrates and *Drosophila*

	Vertebrate	*Drosophila*
Guidance cell	Yes	Yes
Hypoxia sensor	PHD1–3, Hif-1,2	Sima, Togo
Soluble factors	VEGF	FGF
Factors for cellular interaction	Notch/Delta	Notch/Delta

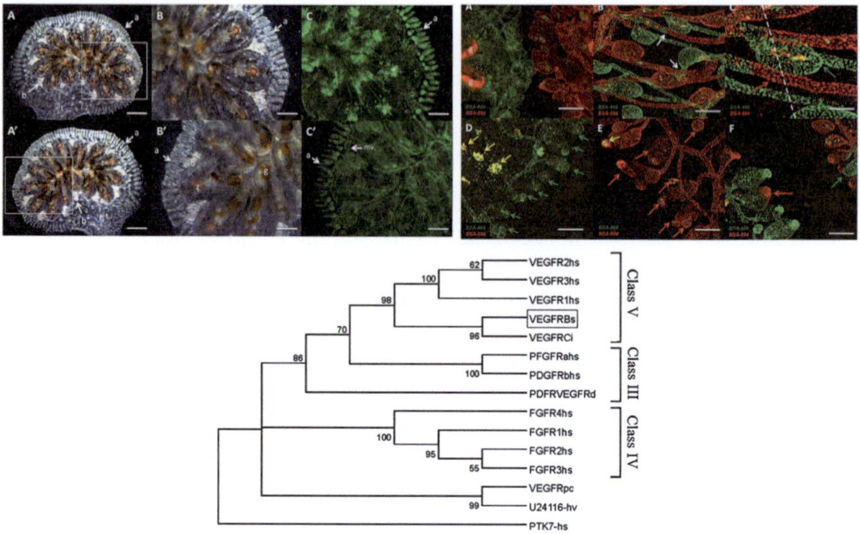

Fig. 3.3 *Botryllus schlosseri*. A colony of Botryllus (*top left*). In *green* the external circulatory system is visible. Fusion of the vascular tubes (*red* and *green*) (*right*). A molecular phylogeny tree (*bottom*). It is seen that the VEGF receptor cluster within class V (from Tiozzo S, Voskoboynik A, FD Brown, De Tomaso AW Developmental Biology 2008, 315 (1):243–255 [96], Permission to reproduce PlosOne)

under the control of inducible factors of hypoxia (HIF) homologs called Sima and Togo. In contrast to vertebrates, where three HIF prolyl hydroxylase are present, only one homologue of prolyl hydroxylase is found in *Drosophila* [94, 95]. One of the Branchless/FGF targets is the Delta/Notch system, and it appears that hypoxia tolerance in *Drosophila* is produced by molecular interactions whose main node is represented by Notch.

An interesting species with regard to the vascular system of invertebrates is *Botryllus schlosseri*. *Botryllus* is a marine invertebrate found abundantly on the coasts of the Atlantic Ocean, and whose zoïds reproduce in an asexual manner. These are fixed on rocks and form "colonies" surrounded by a unique tunic. Antony de Tomaso studied the vascularization of this species in detail [96]. *Botryllus* possesses internal and external circulatory systems. The external circulation consists of branched tubes, which terminate in structures situated at the periphery called ampullae. These vessels allow cells that are not yet well-characterized at the present time to circulate. These cells have defense properties (phagocytosis, etc.) and they appear to be of several cell types. It should be noted that these tubes bud and have the ability to branch. In addition, experiments using two different transgenic *Botryllus* colonies with either a red or green vasculature show that tubes can fuse (Fig. 3.3). It is remarkable that a homologue of the VEGF receptor is expressed in *Botryllus* and this at the level of the myoepithelial cells which, it should be remembered, line the outside wall of the tubes, the internal matrix being in contact with the fluid phase! This receptor (Fig. 3.3), in terms of molecular evolution, is not very distant from the

Table 3.2 Comparison of circulatory systems of different species

Taxonomic class	Type of vascular system	Contractile organ	Presence of an endothelium
Prolifera	No	No	No
Cnidaria	No	No	No
Cephalopodia	Open or closed	Multichambered heart	No
Annelid	Closed	Contractile vessels	No
Arthropodia	Open	Uni-chambered heart	No
Amphioxus	Closed	Contractile heart	No
Hagfish	Closed	Bi-chambered heart	Yes
Osteichthyes	Closed	Bi-chambered heart	Yes
Amphibia	Closed	Tri-chambered heart	Yes
Reptiles	Closed	Tri-chambered heart	Yes
Birds	Closed	Four-chambered heart	Yes
Mammals	Closed	Four-chambered heart	Yes

Modified from Monahan-Earley R., Dvorak A.M., Aird W.C. (2013) J Thromb Haemost 11 Suppl 1, 46–66 [88]

human VEGF receptor. VEGF itself has not been clearly identified, even if a team describes it in the myoepithelial cells themselves, but using a questionable visualization technique [97]. The consequences of these observations and their more general meaning are discussed later (see "Philosophy of the vascular tree").

Table 3.2 summarizes the characteristics of the vascular system in the different species during evolution.

4 Where It All Began

Algire had observed changes in blood vessels when treating fibroblasts with a chemical methyl cholanthrene [98]. This suggests that factors (after stimulation) could be produced from the cells surrounding the blood vessels, such as fibroblasts that stimulate vascular growth.

In 1948, Michaelson [99] published his observations concerning the vascular system of the retina. He had developed a technique, which consisted of injecting India ink into the arterial system, thereby allowing him to visualize the retinal vascularization. Using this method, he studied the development of the retinal vessels and showed that these vessels emerged through the optic nerve to colonize the surface and the interior of the retina. On the basis of these observations, Michaelson postulated that a factor, which he named factor X, was present and induced the budding of new vessels, and that this factor was regulated by hypoxia. Furthermore, in 1951, Campbell [100] observed that the number of capillaries in the retina increased in an oxygen-deficient environment. These observations strongly suggested the presence of mediators of angiogenesis.

However, the answer as to the identity of this factor came from a completely different field, tumor biology. Indeed, tumors are characterized by abundant angiogenesis, and it is the study of tumor angiogenesis that provided the long-awaited answer to the identity of factor(s) X. It is clear that modern research on angiogenesis has really begun with Judah Folkman (Fig. 4.1, Table 2.2) because (1) he provided indisputable experimental evidence that the process of angiogenesis is crucial for in vivo development of tumors and (2) he showed that this was because of a diffusible factor he called "tumor angiogenesis factor" or TAF.

A famous experiment conducted by Michael Gimbrone, a post-doctoral fellow, and by Judah Folkman in 1972 is illustrated in Fig. 4.2 [101]. In this experiment, pieces of a rabbit tumor, called "Brown-Pearce epithelioma," were implanted either at the iris, which is richly vascularized, or in the anterior chamber, a non-vascularized part of the eye. Gimbrone and Folkman observed that the tumors implanted at the iris had an exponential growth whereas the tumors implanted in the avascular region did not grow but nonetheless remained viable. In the words of

A. Bikfalvi, *A Brief History of Blood and Lymphatic Vessels*,
https://doi.org/10.1007/978-3-319-74376-9_4

Fig. 4.1 Judah Folkman, pioneer of the tumor angiogenesis field and discoverer of Tumor angiogenesis factor (TAF). Image courtesy of Harvard University Photo Service courtesy of Harvard Public Affairs and Communication (Jon Chase/Harvard Staff Photographer)

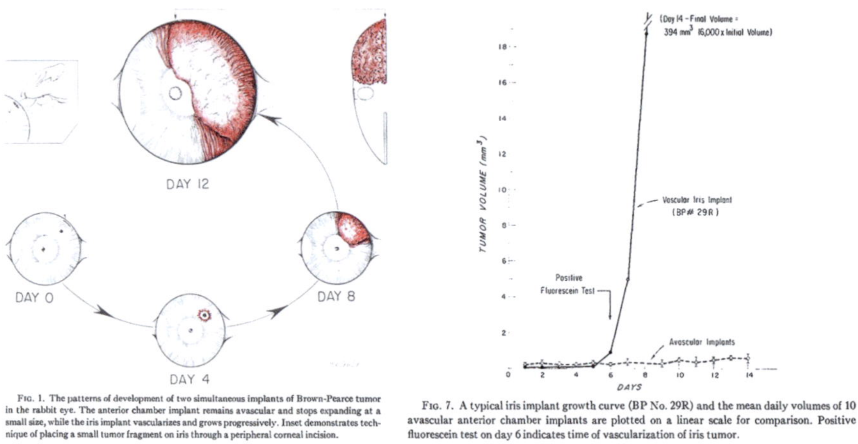

Fig. 4.2 Illustration of the experiments carried out by de Gimbrone and Folkman. Implantation of tumor tissue either at the level of the avascular cornea or near the iris. Implantation technique (*left*). Quantification of tumor growth of these tumors after implantation (*right*). It is seen that tumors implanted near the iris grow exponentially because it is vascularized whereas avascular tumors implanted in the cornea do not develop. (Figure from Gimbrone MA Jr, Leapman SB, Cotran RS, Folkman J (1972) J Exp Med, 136(2):261–276 [101]. Permission to reproduce: Rockefeller University Press)

Michael Gimbrone and Judah Folkman: "*These experiments provide in vivo evidence that prevention of neovascularization can block the growth of a solid tumor at an early stage. It appears that vascularization permits an implant to enter exponential growth, while avascularity forces it to remain dormant at a small size*" and further

"Population dormancy, therefore, appears to be the fate of avascular solid tumors. If neovascularization can be prevented, local malignant growth and perhaps distant metastasis will not occur. These observations thus suggest that specific blockade of tumor-induced angiogenesis would be effective in controlling neoplastic growth".

What has just been written is the fundamental paradigm of angiogenesis and represents the framework in which much research has been conducted in the following decades. However, in the light of current knowledge, this concept/ dogma is in some need of revision. This is discussed later.

A factor called TAF (Tumor angiogenesis factor) was identified by Judah Folkman from the melanoma tumor known as "Walker 256" implanted in rats [102]. Tumors were chemically fractionated and extracts were tested using the method of rat dorsal air sac to show their ability to induce angiogenesis (Fig. 4.3). For Folkman this

ISOLATION OF A TUMOR FACTOR RESPONSIBLE FOR ANGIOGENESIS*

BY JUDAH FOLKMAN, M.D., EZIO MERLER, PH.D., CHARLES ABERNATHY, M.D., AND GRETCHEN WILLIAMS

(From the Departments of Surgery and Pediatrics, Children's Hospital Medical Center and Harvard Medical School, Boston, Massachusetts 02115)

(Received for publication 15 September 1970)

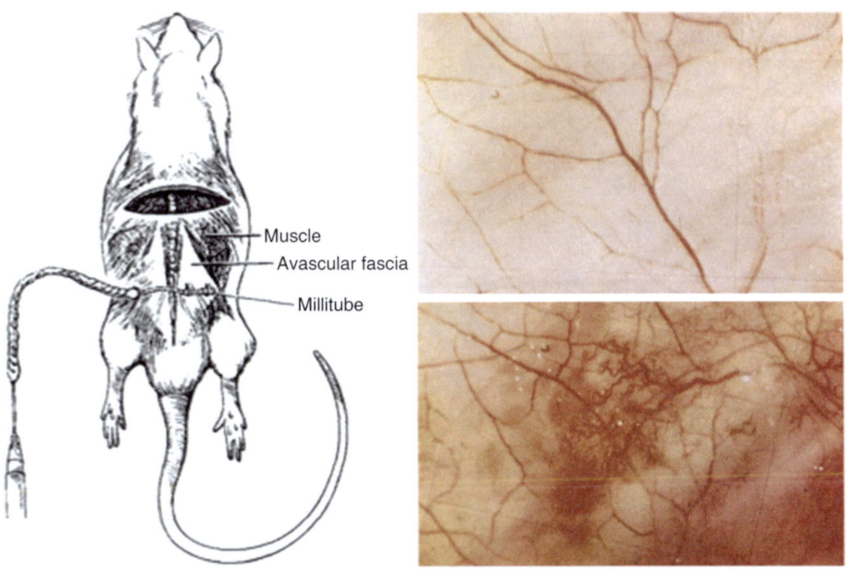

Fig. 4.3 Title of the article by Folkman (*top*) and illustration of the experimental setting that made it possible to evidence the existence of the Tumor angiogenesis factor (TAF) using the dorsal chamber in the rat. The two micrographs on the *right* show that TAF strongly stimulates angiogenesis (*bottom*) compared to control (absence of TAF) (Folkman J, Merler E, Abernathy C, Williams G. Journal of Experimental Medicine 1971, 133(2):275–288 [101]. Permission to reproduce: Rockefeller University Press)

method was more reliable than other tests used at that time, such as using the chorioallantoic membrane of chicken or implantation into the anterior chamber of the eye. Indeed, it was assumed at the time that the angiogenic effects were caused by nonspecific reactions such as tissue irritation or inflammation that appeared in some of the tests. In the same article, Folkman made a hand drawing where he drew the non-growing tumor and a growing vascularized tumor that was induced by the TAF (Fig. 4.4).

In the words of Judah Folkman: "*The presence of a tumor-angiogenesis factor suggests a transfer of information from tumor cells to capillary endothelial cells. The relationship between tumor cells and endothelial cells may be interdependent. Tannock has shown that the nutritional environment of tumor cells becomes poorer as the spacing between blood vessels increases, and this leads to a decreased rate of proliferation and cell death. His work implies that the rate of proliferation of endothelial cells may limit indirectly the rate of tumor growth*".

The pivotal moment that marks the initiation of modern research on angiogenesis is the article published in 1971 in the Journal of Experimental Medicine [102]. Folkman wrote in the introduction: "*The growth of solid neoplasms is always accompanied by neovascularization. This new capillary growth is even more vigorous and continuous than a similar outgrowth of capillary sprouts observed in fresh wounds or in inflammation. Many workers have described the association between growing solid malignant tumors and new vessel growth. However, it has not been appreciated until the past few years that the population of tumor cells and the population of capillary endothelial cells within a neoplasm may constitute a highly-integrated ecosystem.*"

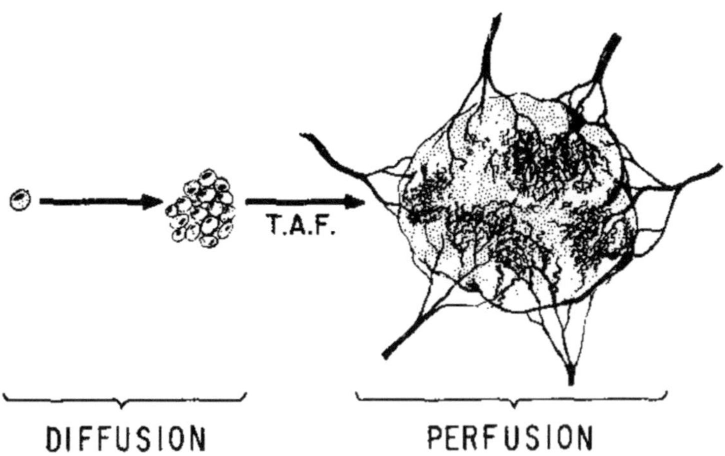

Fig. 4.4 Original drawing by Folkman, from the same article, describing the Tumor angiogenesis factor (TAF). A tumor, having acquired a certain size, has the ability to produce TAF. TAF vascularizes the tumor, which then develops rapidly (Folkman J, Merler E, Abernathy C, Williams G. Journal of Experimental Medicine 1971, 133(2):275–288 [102]. Permission to reproduce: Rockefeller University Press)

It should be noted that Folkman uses the term "highly integrated ecosystem." This implies the existence of intercellular and intracellular signaling systems with mechanisms of amplification and inhibition between different cell populations. This can also be seen as a complete control system with numerous control and feedback loops between tumor cells and blood vessels. The reader cannot escape the view that these interactions and loops may constitute potential therapeutic targets. A more detailed discussion of this notion is provided in the last part of this book.

Folkman formulates the paradigm of angiogenesis in a detailed way in an article published on November 18, 1971 in the New England Journal of Medicine [103]. In his famous article, Folkman put forward very detailed arguments derived from experimental observations, which led him to the following conclusions:

1. Blood vessels in tumors are new, and the tumor actively induces them
2. Induction is caused by tumor-derived factors (TAF) that are diffusible (diffusion means that it can penetrate the tissues to act more remotely)
3. These diffusible proteins actively attract the blood vessels
4. If this process could be blocked or inhibited, tumors should remain in this case of small size, or even regress

This is a research program for several decades! We see in subsequent chapters how this program has been implemented by the different researcher teams. These conclusions are generally correct, but must be modulated or revised in the light of current knowledge.

5 Culture of Vascular Cells In Vitro

Ross Harrison (1870–1959) was the first to attempt to culture tissues in vitro using culture dishes. In 1907 he had grown tadpole tissue in a coagulated interstitial liquid from frog using the hanging drop technique [104]. This technique is still used today to generate spheroids in vitro and to culture them.

A major figure in the development of cell culture was Alexis Carrel, a French surgeon who made his career in the United States and worked at the time at the famous Rockefeller Institute in New York. Basing his work on the technique developed by Harrison, he succeeded in 1912 for the first time in cultivating tissue explants containing animal cells in vitro. These first cultures were established from chicken heart in a medium developed by Carrel containing coagulated plasma, serum, saline, and a chicken embryo extract [105, 106].

The culture conditions were then greatly improved by the development of defined culture media. Harry Eagle (1905–1992) was the first to define more accurately the nutrient requirements for cells grown in vitro. This resulted in the development of the minimal medium called Eagle's Minimal essential medium (MEM) [107]. Subsequently, Renato Dulbecco succeeded in improving the conditions of culture by developing an optimized medium, now famous as DMEM (Dulbecco's modified Eagle medium), which is still used today by most laboratories in the world. Endothelial cells were isolated from various species, including bovine, rodent, and human sources. If the in vitro culture of bovine endothelial cells was easier, it has been more difficult to establish in vitro cultures of human endothelial cells.

Eric Jaffée at Cornell University in New York, and Michael Gimbrone, then a post-doctoral fellow in the Judah Folkman laboratory in Boston, were able to isolate human endothelial cells from newborn umbilical cord and to define exactly the in vitro culture conditions and propagation of these cells [108, 109] (Fig. 5.1). It was the era of "cord hunting" and many laboratories were embarking on this path. In the laboratory where I worked during my thesis, we were in contact with different maternity hospitals in Paris and there were fixed days to collect the umbilical cords. Beakers were filled with culture medium in which the obstetricians then placed the umbilical cords of newborns, which we then collected the next day.

A. Bikfalvi, *A Brief History of Blood and Lymphatic Vessels*,
https://doi.org/10.1007/978-3-319-74376-9_5

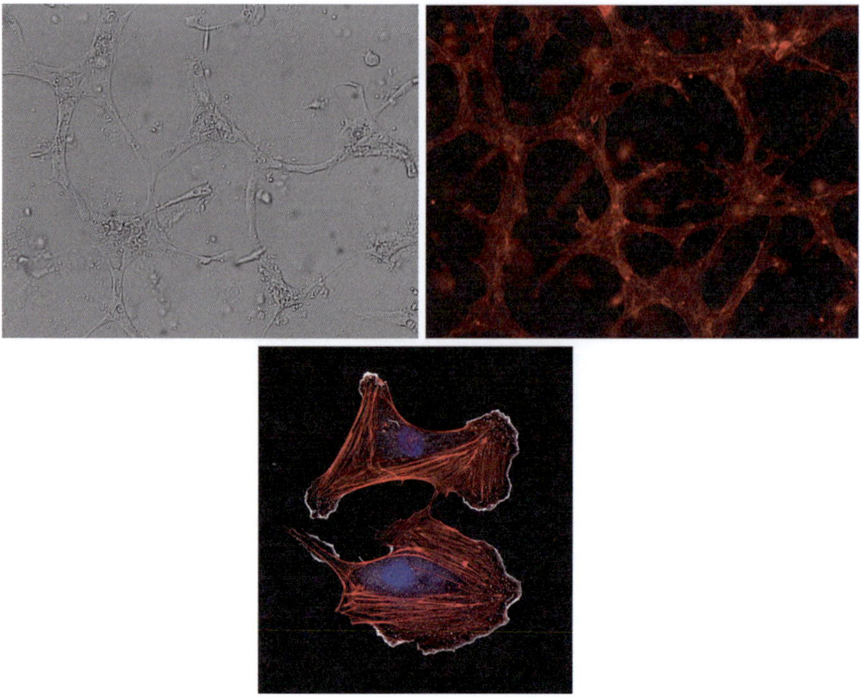

Fig. 5.1 Endothelial cells in culture. Endothelial cells cultured from the umbilical cord are represented in a vitro angiogenesis system (*top*). Phase contrast image (*top left*). Fluorescence image (phallodin in *red*) (*top right*). Endothelial tube formation can be evidenced under these conditions. Microvascular endothelial cells of the lung after immunostaining using an antibody against cortactin (*white*) and actin (*red*) (*bottom*). The nuclei are marked in *blue*. The arrangement of the cytoskeleton and the actin fibers is clearly seen. Images from the author's team (L. Cooley and T. Daubon post-doctoral fellows in the laboratory)

The isolation of human capillary endothelial cells was a more difficult task. Although it was fairly easy to obtain these cells from animal sources (especially from the adrenal cortex in bovine), human capillary endothelial cells were difficult to grow in vitro because their need for nutrient and growth factors was more important. Different sources were then tried, such as the newly circumcised newborn foreskin or omental fat (fat tissue in the abdominal cavity). These difficulties were related to the fact that the culture conditions of these cells were not yet optimized because the factors responsible for vascular growth had still to be identified.

Endothelial cells, similar to any other cell type, are grown in plastic dishes (Petri dishes) in a defined medium containing amino acids, glucose, vitamins, and animal or human serum. The acid or basic character measured by the pH must also be regulated and therefore bicarbonates, which are important for pH stabilization, are required. All this, of course, must be placed in a 37 °C incubator. The tissue extracts are then added to optimize the growth of endothelial cells, in particular brain extracts

Table 5.1 Markers of blood vascular and lymphatic endothelial cells

	Vascular endothelial cells	Lymphatic endothelial cells
CD31	++	+
VE-Cadherin	++	−
PAL-E	++	−
Weibel–Palade bodies	++	−
Willebrand factor	++	−
CD34	++	−
Lyve-1	−	++
Podoplanin	−	++
VEGFR2	++	+−
VEGFR3	+	++

presumed to be rich in growth factors. An extract frequently used and supplied by approved companies was the ECGS or "endothelial cell growth supplement."

Another aspect is the use of proteins called matrix proteins to facilitate adhesion, migration, and growth of endothelial cells. These proteins are also part of what the general public knows as "connective tissue." Collagen, for example, is part of this and is used precisely in its denatured form for the culture of endothelial cells in order to promote the maintenance of these cells in vitro. Endothelial cell culture is now much better defined and extracts such as Endothelial cell growth supplement (ECGS) have been replaced by vascular growth factors.

What about the culture of other vascular cells? There are not only endothelial cells in a vessel but also pericytes in the capillaries and smooth muscle cells in large vessels. These cells can also be isolated and cultured in vitro and various techniques have been developed to isolate both pericytes and smooth muscle cells [110–112].

Nowadays the situation has become much easier (but more expensive!) because many companies offer frozen vascular cells (Lonza, Promocell, ATCC), which can therefore be purchased. This avoids the step of isolation (and characterization), which is time consuming.

What are the morphological and molecular characteristics that allow the identification of vascular wall (Table 5.1). For endothelial cells, because they are in direct contact with blood, the surface must be non-thrombogenic, which means that blood coagulation is inhibited. This is caused by exposure to the surface of certain phospholipids that prevent activation. In addition, endothelial cells harbor special structures called Weibel–Palade bodies, named after their discoverers, Ewald R. Weibel and George Emil Palade in 1964 [113]. These bodies appear under the electron microscope as striated rods and contain an important molecule which intervenes in the interaction with blood platelets called von Willebrand factor. On the surface of endothelial cells, some characteristic markers are present, the best known being Cluster of differentation 31 (CD31), also called Platelet endothelial cell adhesion molecule (PECAM-1). This marker is used in general to identify endothelial cells under a microscope after immunolabeling. Other markers include

Pathologische Anatomie Leiden-Endothelium (PAL-E) and Cluster of differentiation 34 (CD34). The VEGFR2 receptor can also be used as a marker because it is present on the surface of endothelial cells.

For lymphatic endothelial cells, markers are different, the best known being the Lymphatic Vessel Endothelial Receptor 1 (Lyve-1) marker. Lyve-1 is the receptor of a molecule called hyaluronic acid. Hyaluronic acid is a sugar of a particular species called glycosaminoglycan. This sugar is present in the extracellular fluid, plasma, and lymphatic vessels. It is assumed that hyaluronic acid uses Lyve-1 to enter the lymphatic circulation. Another important lymphatic marker is podoplanin, also present in lymphatic vessels.

It should be noted that not all of these markers are absolutely specific. CD31, for example, is detected in vascular endothelial cells and in a fraction of lymph endothelial cells. In addition, monocytes/macrophages also express CD31. These markers are, however, very useful because, if combined with other markers, they are of great help for the identification of the endothelial cells.

Concerning markers of pericytes and smooth muscle cells, these are represented by desmin, actin α2 smooth muscle cells (α2SM actin) or neural/glial antigen-2 (NG2). These molecules can be easily detected using specific antibodies against these molecules.

Discovery of the First Stimulating Factors of Blood Vessels

6

The work of Judah Folkman indicated that one or more factors produced by a tumor could stimulate angiogenesis. These factors were called Tumor Angiogenesis Factors (TAFs), and various laboratories set out to identify these TAFs. The problem was that even if the culture medium or the extracts of the tumor cells or of other tissues showed an important stimulatory activity for endothelial cells, the responsible factors could not be efficiently purified. This changed with the discovery that these factors had a strong affinity for heparin. Using heparin-sepharose chromatography, Michael Klagsbrun and Jay Shin from the Judah Folkman laboratory, and researchers from several other laboratories, were able to isolate these factors—which were called heparin-binding growth factors—and to determine their amino acid sequences [114–116] (Fig. 6.1).

Chromatography is a method of separating proteins using columns containing a sepharose gel at various densities, on which a tissue extract or a culture medium to be separated is deposited. In the case of heparin-sepharose, which has a high capacity for binding to basic proteins, heparin is fixed irreversibly to sepharose. A chromatography column is then filled with the functionalized sepharose. The heparin-binding proteins are thus retained in the column in a first step. In a second step, these proteins are then detached from the column (eluted) by a buffer with a high ionic strength (elution with high concentration of salt) and collected.

Many teams then began to purify these factors from many normal or pathological tissues. This was the case with Denis Gospodarowicz, Roger Guillemin (winner of the Nobel Prize for Medicine), and Peter Böhlen, all localized at the west coast of the United States. These factors were first called according to the organ from which they were identified. For example, Yves Courtois's team, he being, at that time, director of an eye research laboratory in Paris, had identified in the eye a factor called eye-derived growth factor. It was therefore important to reach a consensus and to establish an appropriate nomenclature. This was first done in 1985, and these factors were called acidic or basic Fibroblast Growth Factors (FGF) according to their amino acid sequences [117] and then named by assigning them a number for the factor type. Thus, acidic FGF was designated FGF-1 and basic FGF-2.

A. Bikfalvi, *A Brief History of Blood and Lymphatic Vessels*,
https://doi.org/10.1007/978-3-319-74376-9_6

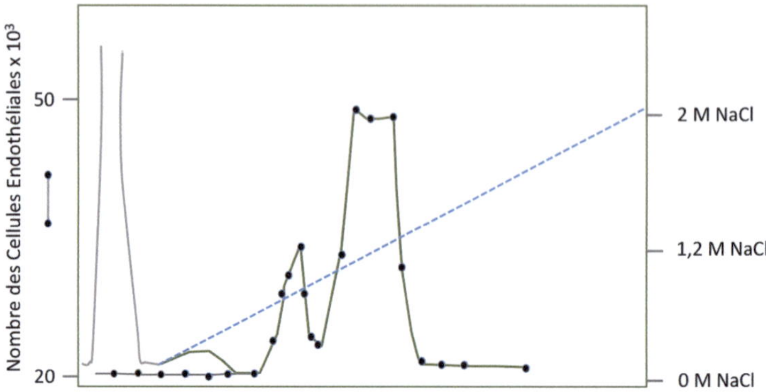

Fig. 6.1 Isolation of endothelial cell growth factors by heparin-sepharose chromatography. The tissue extract is homogenized and deposited on a heparin-sepharose column. A gradient containing increasing concentrations of NaCl is then applied and elution is performed. The various fractions eluted are then tested for their ability to stimulate the proliferation of endothelial cells. It can be seen that the most stimulatory fractions are eluted from 1.2 M NaCl. (Figure redrawn and modified by the author of Klagsbrun and Shing, Proc Natl Acad Sci USA, 1985, 82: 805–809) [121]

The list of these factors has grown considerably, now comprising 23 members (FGF-1 to FGF-23).

However, were FGFs the long sought-after TAFs? I still remember the animated discussions at a scientific meeting in Paris in 1995, with Werner Risau, director at the time of the Max Planck Institute on angiogenesis in Bad Nauheim, Germany. He was barely 40 years old and was then one of the most respected scientists in the field. One of the articles he published was entitled "What if anything is an angiogenic factor?" [118] to relativize the fact that any supposed factor may represent an angiogenic factor, even if it stimulated the development of blood vessels when applied externally. He always said that the expression profile of FGF did not fit with a role in vascular development and that FGF (at least some of them) did not have a so-called "signal" sequence that allows them to be efficiently exported from the inside of the cell into the extracellular medium, which is a property absolutely required for an angiogenic factor. The feeling shared by many researchers at that time was that FGFs were not the long sought-after TAFs.

If the FGF was not the TAF, as it appeared at the time, what then was the biochemical nature of this activity? Many teams continued their research, including the team of Judah Folkman.

In Boston, Judah Folkman contacted the famous Boston chemist Bert Vallée and obtained a very important funding to identify this factor. Bert Vallée was a celebrity in Boston and a highly respected chemist. Nevertheless, this collaboration was not fruitful because Bert Vallée was determined to pursue this research all by himself. A molecule called angiogenin was isolated and characterized [119, 120]. However, this molecule proved to be a ribonuclease. It had limited activity on vascular cell proliferation and lacked an adequate temporo-spatial expression to attribute angiogenin to a significant role in tumor angiogenesis. Exit therefore angiogenin.

In the 1990s, the idea emerged that there must exist vascular-specific factors that had the ability only to stimulate angiogenesis. This(these) factor(s) would have a role in embryonic development and during vascularization of tumors and other pathologies. The era of Vascular Endothelial Growth Factor (VEGF) was about to begin!

7 Vascular Endothelial Growth Factor: The Cornerstone of Vascular Development Factors

In 1986, Napoleone Ferrara and Jean Plouet (Fig. 7.1), both postdoctoral fellows in Denis Gospodarowicz's laboratory in San Francisco, detected a stimulating activity of endothelial cells in the culture medium of bovine and mouse pituitary cells that was eluted from a heparin-sepharose column at much lower ionic strength than fibroblast growth factors [122, 123].

Even more remarkable is that this activity had a very high specificity to stimulate endothelial cells and did not, it seems, stimulate other cell types. Again, heparin —a molecule important not only because of these various biological roles but also from a technological development point of view—was at the basis of this discovery. This factor was called Vascular Endothelial Growth Factor (VEGF). The surprise came from the structure of this factor, which is at every point identical to an already identified factor, called Vascular Permeability Factor, also known as VPF. This factor had been identified a few years earlier by Harold Dvorak, director of a laboratory at the Dana Farber Cancer Institute in Boston, who was working on making vessels more permeable [124]. Vascular permeability plays an important role in inflammatory states and in cancer. Nevertheless, the team at Dana Farber had not studied the VPF for its properties to stimulate angiogenesis.

I was in contact with Jean Plouet as soon as he returned from the United States and we have done some work together on the characterization of VEGF receptors on endothelial cells. The scientific community very quickly focused its attention on VEGF and many teams entered the race. The team of Werner Risau (Max Planck Institute, Bad Nauheim, Germany), already mentioned above, centered all efforts on VEGF at that time. A young and brilliant researcher, Peter Carmeliet, from the University of Louvain, also joined the competition (Fig. 7.2). He was the head of a large team that was extremely well-funded, thanks to patents obtained for Tissue Plasminogen Activator (tPA), by Désiré Collen, director of the Institute at the time Peter Carmeliet was working. This factor (tPA) is important in the lysis of the clot (fibrinolysis). Peter Carmeliet used primarily innovative genetic approaches to invalidate a given gene in the mouse or to overexpress it. He systematically applied

A. Bikfalvi, *A Brief History of Blood and Lymphatic Vessels*,
https://doi.org/10.1007/978-3-319-74376-9_7

Fig. 7.1 Napoleone Ferrara (*left*) and Jean Plouet (*right*), the discoverers of VEGF. Napoleone Ferrara graciously provided his photo. Picture of Jean Plouet: permission to reproduce Cancer Research

Fig. 7.2 Peter Carmeliet, the genetic approach to the study of angiogenesis. Image courtesy of Peter Carmeliet

this technology, also called transgenesis, to the various factors involved in angiogenesis in order to characterize their role in vivo in an entire organism.

Meanwhile, Napoleone Ferrara had joined Genentech in San Francisco, California, and was using similar approaches combined with molecular biology and advanced biochemistry. Genentech was a start-up at the time and therapeutic development in the area of angiogenesis was of definite interest to the society. Napoleone Ferrara's joining the society was extremely beneficial for the development of the first anti-angiogenic drug used clinically, as we see later. This was the beginning of the "VEGF period" of angiogenesis and many results were obtained. VEGF receptors were identified and their structures determined [125–127]. These receptors were tyrosine kinase receptors, the activation of which induced phosphorylation (inclusion of a phosphate group at the amino acid tyrosine) and which had three members : VEGF receptor-1 (VEGFR1), VEGF receptor-2 (VEGFR2), and VEGF receptor-3 (VEGFR3). Signaling pathways have been elucidated, including the involvement of three important pathways, mitogen activating protein (MAP) kinases, PI3 kinases, and protein kinase C (see Section 10.7 "Signaling Induced by Angiogenic Factors") [128].

The genetic invalidation of VEGF or its receptors revealed an obvious vascular phenotype [125, 129, 130]. Without the presence of VEGF, vascularization did not develop in the embryo. The inactivation of a single allele already had negative effects on vascular development. VEGF was therefore identified as the primary factor for the regulation of angiogenesis. Furthermore, it was shown that this factor was very important in tumor angiogenesis because its inhibition blocked tumor growth [125, 131]. It was generally accepted at that time that a serious scientist had to work on this molecule.

Nevertheless, some scientists (including us) did not believe in the omnipotence of VEGF and continued to look in other directions. The difficulty of finding alternative paths came from the following fact. Many molecules involved in other biological processes are able to stimulate the growth of blood vessels at least in vitro, and this was already evident at the time. Are they, however, factors that can intervene in angiogenesis in physiology or pathology? As we have said before, the late Werner Risau entitled one of his articles "What, if anything, is an angiogenic factor?" [118] to evoke the plethora of factors already described at the time. The problem was to differentiate potential (hypothetical) factors from "real" factors with a role in vascular development in physiology or pathology. How to answer this question objectively? If an invalidation of a gene, as can be done in mice, clearly shows abnormalities in vascularization, this indicates that this gene is important. This is precisely the case for VEGF, because the loss of a single allele already shows severe vascular defects. Nevertheless, the absence of obvious signs of anomalies of vascularization during embryonic development does not exclude a gene from playing a role in angiogenesis because of functional redundancy with other factors whose expression can compensate for the loss of this gene. It was, therefore, necessary to stimulate the imagination to demonstrate convincingly the true role of a molecule and its specificity. We deal with this a little later in this book and in more detail.

8 Inhibition of Angiogenesis, "Disappointments and Success"

Previous work by Judah Folkman had indicated that abnormal vascularization could be inhibited and that this could have a beneficial effect in pathological situations where vascularization was excessive. Many pharmacological substances are indeed capable of inhibiting the growth of endothelial cells, including synthetic molecules, plant extracts, and snake venoms.

Nevertheless, Folkman postulated the existence of endogenous inhibitors, that is, inhibitors that, in the physiological state or in pathology, contribute to limiting angiogenesis [132] and to ensuring quiescence of the blood vessels ("dormant state"). This idea came from the fact that tumor ecology calls on stimulating and inhibitory factors that are in a state of shifting equilibrium (Fig. 8.1). Michael O'Reilly, a post-doctoral fellow in Folkman's laboratory, set out to identify these factors. He had noticed that Lewis lung carcinoma tumors implanted in mice had much greater dissemination after being removed. He therefore postulated the existence of inhibitory molecules originating from the tumor that are capable of inhibiting the development of blood vessels. Two molecules were identified and were called "angiostatin" and "endostatin" [133, 134]. Angiostatin is a fragment of plasminogen, an enzyme involved in the degradation of fibrin that forms during blood clotting. Fibrin also accumulates in a tumor, between the vessels and the tumor cells, and constitutes what is called a temporary matrix. Endostatin is a fragment of collagen 18, an important component of the extracellular matrix. Endostatin, according to the authors, had the effect of rendering tumors quiescent after repeated administration. The announcement of these results published in prestigious journals such as *Cell*, *Nature*, and *Science* was a bombshell in the scientific community [128, 129]. From the *New York Times* to the French newspaper *Le Monde*, the front pages of these journals and newspapers reported a spectacular advance in cancer research [135]. Jim Watson, discoverer of the structure of DNA, exclaimed that cancer would be definitively defeated in a few years!

After the enthusiasm came the disappointment. Many laboratories had great difficulty reproducing the results of the Boston team and could not demonstrate an inhibitory activity, and this even in vitro on endothelial cells in culture. At the time,

A. Bikfalvi, *A Brief History of Blood and Lymphatic Vessels*,
https://doi.org/10.1007/978-3-319-74376-9_8

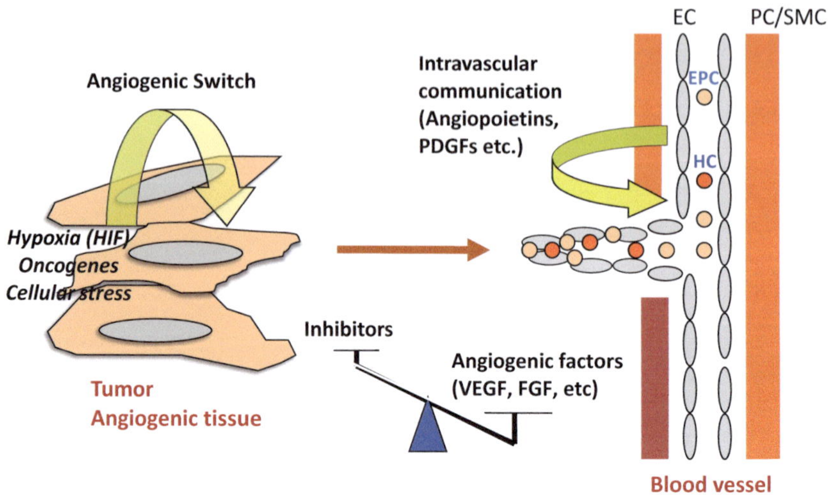

Fig. 8.1 Cellular and molecular interactions in tumor angiogenesis. The tumor cell or "angiogenic" tissue emits factors that stimulate vessel growth. These factors are induced in cells (tumor cells or "angiogenic" cells in other pathological tissues) by a mechanism called the angiogenic switch. This mechanism is not fully understood, but involves regulators of hypoxia, cell stress factors, and oncogenes. These factors then diffuse and stimulate the growth of endothelial cells. There are natural inhibitors that normally counterbalance the effect of angiogenic factors. The angiogenic factors are, nevertheless, in excess of the inhibitors, which leads to the stimulation of angiogenesis. There are also so-called "guidance" factors that are involved in the directional growth of vascular tube. Angiogenic factors then activate the receptors on the surface of the endothelial cells and cause sprouting. In addition, there is an accumulation of hematopoietic cells from the bone marrow that enhance angiogenesis through the secretion of additional molecules. The final process consists of an attraction ("recruitment") of pericytes to the vessels by attractive factors represented by PDGFs and angiopoietins. Nevertheless, recruitment of pericytes is incomplete in the case of cancer, and these cells have only a loose attachment to the vascular tube. This is different from normal vessels because, in this case, the attachment of pericytes is very tight. *EC* endothelial cells, *PC* pericytes, *SMC* smooth muscle cells. Figure of the author

we tested preparations sent by Ruppert Timpl from the Max Planck Institute in Munich, the undisputed authority of matrix proteins and collagens. I was initially enthusiastic because the post-doctoral student working in my laboratory on the subject had demonstrated an inhibitory activity on the proliferation and migration of endothelial cells. Nevertheless, after a reexamination of the experimental conditions, the effect demonstrated was simply caused by the solution in which endostatin was dissolved and not by the molecule itself. This effect is called a "buffer effect" of which one must always be aware.

In addition, in vivo injection, which had shown a dramatic effect in mice, posed problems because O'Reilly had injected endostatin precipitates because of poor solubility. On the other hand, it was unlikely that a molecule, after repeated injection, suddenly abolished tumor growth. A commission of the National Institutes of Health (NIH) was set up to visit Judah Folkman's laboratory to detect possible fraud regarding these results, but no wrong doings were detected. The problem of the non-reproducibility of endostatin was finally partially solved because it was realized that the molecule was very unstable, even transported in a box containing dry ice. Rumor circulated in the scientific community that Judah Folkman was testing the stability by putting endostatin in a box containing dry ice and driving it around Boston in a taxi to test the activity of the molecule after transport! A mutant of this molecule was manufactured which, in principle, retained its activity. Nevertheless, the experiment of "quiescence" in the mouse has never been reproduced.

Other molecules representing fragments of matrix proteins have been discovered, such as "tumstatin," fragment of the NCI domain of the α3 chain of collagen IV or "canstatin," fragment of the C-terminal global domain of the α2 chain of collagen IV. Ragu Kalhuri of the Dana Farber Institute in Boston had shown that tumstatin had inhibitory effects on angiogenesis [136]. In mice, the absence of tumstatin resulted in the early onset of tumors. The mechanism of a large number of proteins in this family involves interaction with membrane receptors called "integrins". However, this research has not been pursued and has not led to therapeutic development.

A family of molecules widely studied is the family of thrombospondins. Discovered in 1971 by Nancy Baezinger in blood platelets, these molecules belong to the larger family of matrix proteins [137]. There are two main types of thrombospondin: Thrombosponin-1 (TSP-1) and Thrombospondin-2 (TSP-2). These molecules are expressed in many tissues and also overexpressed in different types of cancer. They have the ability to inhibit angiogenesis, and during tumor progression their expression is lost. In contrast, expression of TSP-1 or TSP-2 in cancer cells inhibited tumor development. Thrombospondins are matricellular proteins that are deposited by cells into the extracellular matrix. They alone are not capable of forming a functional matrix and must therefore interact with other matrix molecules. These proteins can have extremely complex roles and, in some cases, promote tumor development (see Section 16.2 "Inhibition of Angiogenesis"). The cellular partners with which TSP interacts are very diverse and include molecules designated as CD36, integrins, and CD47.

The approach using endogenous inhibitors derived from extracellular matrix proteins as a therapeutic tool has not been successful, at least to date. The pharmaceutical industry has not invested more effort in the drug development of these molecules. Only China has authorized the marketing of a stabilized form of endostatin and a number of clinical trials have been completed or are ongoing (https://clinicaltrials.gov/ct2/results?cond=&term=endostatin&cntry=&state=&city=&dist=).

The most promising strategy that definitively validated the anti-angiogenesis approach was the development of specific inhibitors against VEGF and its receptors. Napoleone Ferrara, then working at Genentech, invested all his efforts in the development of a humanized monoclonal antibody capable of blocking the activity of VEGF [138]. Napoleone Ferrara showed that this antibody was able to inhibit angiogenesis and growth of a large number of tumors in preclinical models. Based on this work, a humanized antibody was developed and tested in clinical studies. The Hurwitz clinical study in 2004 showed for the first time that this antibody was effective in metastatic colon cancer, at a stage of the disease where any other therapy fails [139]. The antibody was granted marketing authorization in 2005 under the name Avastin (see below). Napoleone Ferrara received in 2010 the famous Lasker Prize, which is the forerunner of the Nobel Prize for his work on VEGF and for the development of an effective anti-angiogenic therapy [140].

Other inhibitors of the VEGF pathway were also developed. These inhibitors bind to the so-called "tyrosine kinase" domains of VEGF receptors, which are essential for transmitting the signal within the cell after activation (see Section 10.7 "Signaling Induced by Angiogenic Factors" for detailed explanations). To this end, they must penetrate into the cell to attach to the so-called "tyrosine kinase" domain. These inhibitors have more or less narrow specificities and can also affect, besides VEGF receptors, other tyrosine kinase receptors, although with a lower affinity. Sunitinib or sorafenib are prototypes of these inhibitors. These molecules, especially sunitinib, show in particular an efficacy in clear cell kidney cancer (ccRCC), which is the most common type of kidney cancer.

Stimulating Angiogenesis

9

Cardiovascular diseases (myocardial infarction, cerebral ischemia, or lower-limb arterial disease) contribute significantly to mortality and morbidity in our societies. These diseases are the result of arterial obstruction that consequently reduces the blood supply to the heart or to the peripheral tissues. This results in ischemia or necrosis of the tissue in question. New therapeutic approaches must therefore be developed to treat these diseases. The idea of promoting vascularization in tissues that suffer from a lack of blood supply, as is the case in myocardial infarction, stroke, and lower-limb arterial disease, is a sound strategy. The idea was to promote collateral circulation to save poorly irrigated tissue that suffers from a lack of oxygen supply ("ischemic tissue"). In the chapter on the history of angiogenesis we saw that John Hunter had already somehow anticipated this idea (Chap. 2 "History of the Vascular System").

Preclinical models have shown that the administration of angiogenic factors such as vascular endothelial growth factor or Fibroblast growth factor (FGF) 1, 2, 4 may induce the development of collateral circulation. Experiments using either recombinant proteins or plasmid DNA have therefore been initiated in humans [141, 142]. Initially, observations in favor of an improvement in clinical symptoms in lower-limb arteriopathy confirmed this postulate [143]. One of the first clinical findings was a study of 180 patients (Traffic Trial) to whom FGF2 was administered in recombinant form [143]. This study showed a benefit with an increase in walking distance. In addition, studies using a gene therapy approach by intramuscular administration of a plasmid encoding FGF1 showed some benefits with a twofold reduction in the risk of amputation [142]. However, wider controlled clinical trials have not shown any benefit for approaches using recombinant proteins or plasmid vectors administered to patients [144]. This is therefore a very disappointing result and shows that preclinical observations do not always go hand in hand with what happens in man.

Another strategy to stimulate angiogenesis is the injection of cells from the bone marrow. This approach has shown its efficacy in vivo in preclinical models and in humans in many studies [145]. The initial idea was that the injected cells could be incorporated into the vessels in formation and thus stimulate their growth.

A. Bikfalvi, *A Brief History of Blood and Lymphatic Vessels*,
https://doi.org/10.1007/978-3-319-74376-9_9

Nevertheless, it seems clear today that the pro-angiogenic effect is rather caused by the release of stimulating factors by the injected cells. There are generally two different types of applications. The first is based on the use of Endothelial Cell Progenitor Cells (EPCs, see below) and the second on the use of cells called myeloid cells (cells that give white blood cells). It has been reported that the use of myeloid cells may increase the risk of atherosclerosis by the formation of cells called foam cells that have an accumulation of lipids in their cytoplasm [146]. These cells are packed into the vessel wall and cause thickening and hardening of the vascular wall. According to these data, it would therefore be preferable to use the EPCs and not the myeloid cells in a therapeutic application.

10 The Situation Is More Complex Than Anticipated

Vessels are not uniform tubes but are subdivided into arterial, venous, and lymphatic vessels. Large vessels (arteries, veins, large lymphatic trunks), medium-size vessels (arterioles, venules, medium-sized lymphatics), and capillaries are also distinguished. As we have just seen, the identification of the VEGF family had a considerable impact on both basic and clinical research (see below). Nevertheless, the simple picture that we had following Folkman's work has become considerably more complicated and many modifications/corrections have been made at the cellular and molecular levels.

10.1 TIP Cells

Any developing tubular structure must be oriented in space and follow stimulating or repulsive factors. The blood vessels resemble axons of the nerve cells, which orient themselves and migrate over significant distances to reach their target. Thus, in a vessel, we can hypothesize that specialized cells exist that are responsible for guidance of the vascular tube. These cells were identified by Christer Betzholtz (Fig. 10.1) and Holger Gerhard of the Karolinska Institute in Stockholm and called "tip cells" [147] (Fig. 10.2). These cells are located at the end of blood vessel and emit filopodia that act as extracellular signal–transmitting sensors. Tip cells have no lumen to conduct blood and are followed by cells called stalk cells that have a lumen.

Much work has been done in recent years to clarify the role of these tip cells and their interactions with stalk cells. These cells have been shown to be highly sensitive to VEGF and to follow a VEGF gradient. It also was postulated until recently that VEGFR2 plays the main role in vessel guiding [148]. However, Ralf Adams and his team (Max Planck Institute in Münster, Germany) proposed that the main receptor in tip cells is not the R2 receptor but the VEGFR3, which is overactivated in tip cells [149]. However, this has been contradicted by a recent publication [150] (see later in this book). Thus, it seems that VEGFR2 is still the principal VEGF receptor localized at tip cells and is responsible for sprouting.

A. Bikfalvi, *A Brief History of Blood and Lymphatic Vessels*,
https://doi.org/10.1007/978-3-319-74376-9_10

Fig. 10.1 Christer Betsholtz and the identification of tip cells and of the role of pericytes in angiogenesis. Image kindly made available by Christer Betsholtz

The discovery of the tip cells that guide the vasculature stems from neurobiology in analogy to axonal growth. In 1890, Ramon y Cajal [151] described a structure he named the "axonal growth cone" at the extremity of an axon. Axonal growth cones have filopodia and lammelipodia protrusions, which are important for sensing guidance cues. In analogy, this concept has been transposed to blood capillaries in the following way. Sprouting blood vessels are composed of cells, which, similar to growth cones, lead the vascular tube by sensing environmental cues and thus guiding the vascular tube in a specific direction.

Both growth cones and tip cells exhibit cellular structures that direct cell movement and are linked to the actin cytoskeleton machinery. The difference between axon growth cones and tip cells lies in the fact that tip cells are cellular structures located atop a growing nascent vessel and thus guide a multicellular unit comprising the tip cell and stalk cells that follow the tip cells. Growth cones, in contrast, represent specific structures at the end of axons. However, both sense the microenvironment for guidance cues to move forward.

An intriguing observation was recently made that two endothelial cells constitute the tip of a nascent blood vessel [152]. Both cells extend filipodia and may also be involved in lumen formation through cord-hollowing. Tip cells are particular cells that exhibit specific morphological, phenotypic, and molecular characteristics. Tip cells respond to molecular gradients such as VEGF and VEGFR2 is localized at the filipodia. Furthermore, tip cells and stalk cells are integrated into a molecular circuit that involves VEGFR2, Notch, Delta4, and Plexin D1 [153]. Indeed, VEGF activates VEGFR2 in tip cells, which in turn increase DLL4 and Plexin D1. DLL4 then interacts with Notch1 on stalk cells, decreasing VEGFR2 and thus exerting an inhibitory signal. Additional signaling mechanisms and guidance cues have been identified, such as endothelial cell–derived sema3A, which exerts a repellent function on tip cell filipodia [154]. Finally, tip cells have specific metabolic regulation whereby glycolysis through PFK3B is significantly increased at the level of tip cells in the filipodia [155]. In tumors, PFKFB3 upregulation leads to a more activated endothelium and a dysfunctional vasculature, which can be normalized by PFKFB3 blockade (metabolism is discussed later in this book).

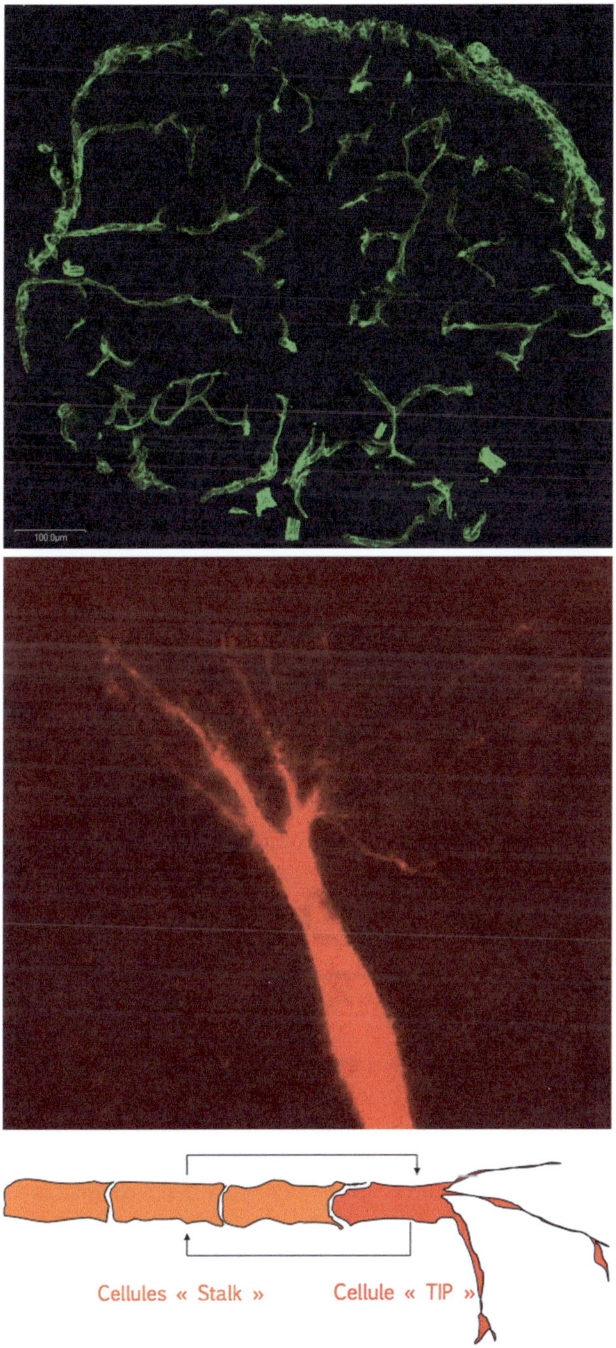

Fig. 10.2 "Tip" and "stalk" cells. Transverse section of a mouse spinal cord (*top*). Blood vessels comprising stalk cells and tip cells contain many tubular structures. A tip cell that emits filopodia (*middle*). Schematic representation of a vascular tube (*bottom*). Pictures and drawing by the author

It has been shown more recently through mathematical modeling and live-imaging approaches that the rate of tip cell selection determines the length of linear sprout extension [156]. Thus, tip cells not only determine the direction of the nascent sprout but also are critical for sprout extension.

This indicates that the tip cell concept imported into vascular biology from neurobiology in analogy to neuronal growth cones constitutes a fertile conceptual framework that has led to a significant body of research that has enriched our knowledge of the vascular morphogenic processes and may have significance for pathology as well.

10.2 Formation of Vascular Lumen

A fundamental step is the formation of lumen in the vascular tube, necessary to convey blood to the tissues, which we now discuss. Another important step is the contact between the different vascular cells in the newly formed vessels and the establishment of blood flow. The mechanism of this process has not yet been elucidated.

The formation of the lumen remains little studied. Different mechanisms have been proposed for lumen formation in epithelial tissues and vessels [157] (Fig. 10.3). These mechanisms are called chord hollowing, cell hollowing, invagination of the plasma membrane (invagination), and multicellular cavitation (cavitation). In the case of blood vessels, three mechanisms were mainly proposed: "cord hollowing," "cell hollowing," and "invagination" [158–161]. Initially, cellular cavitation was thought to be the predominant mechanism. In cellular cavitation, vacuoles inside endothelial cells form and gradually fuse to cross the cells completely. These vacuoles are generated inside the cell or by pinocytosis, that is, by internalization of vacuoles from the plasma membrane. However, several studies have been carried out to demonstrate that tubular cavitation and invagination are the principal and even the only mechanisms of lumen formation (Fig. 10.3) [161, 162].

Tubular cavitation involves two adjacent cells and involves a number of molecular steps whose precise course of action begins to be understood better. It involves the use of adhesion molecules on the surface of endothelial cells called VE-cadherin and N-cadherin, responsible for the close contact between two endothelial cells. This contact is then broken by molecules called "de-adhesion factors." These factors include glycoproteins (CD34-Sialomucins including CD34 and Podocalyxin [PODXL]) and are responsible for the elimination of adhesion molecules from the cell surface. Actin is a contractile molecule that, complexed with myosin, is found inside endothelial cells. This complex is recruited at the interface between two endothelial cells whose contraction initiates the opening of a space between the two endothelial cells. To do this, specific proteins called ERM (E for Ezrin, R for Radixin, and M for Moesin) are phosphorylated, allowing the connection between actin and surface glycoproteins. Reduced actin polymerization leads to destabilized endothelial junctions and consequently to failure in blood vessel lumenization and lumen instability. Recent findings indicate the importance of the actin nucleator and

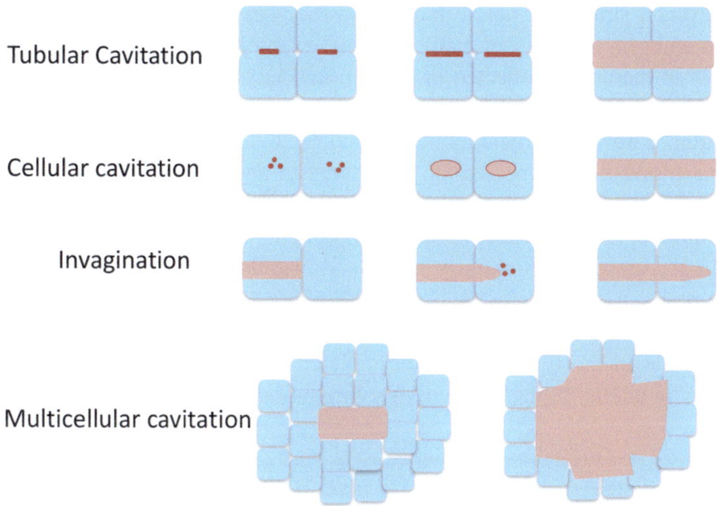

Fig. 10.3 The different mechanisms proposed for the formation of the lumen. Redesigned and modified figure from Sigurbjornsdottir S, Mathew R, Leptin M (2014) Nat Rev Mol Cell Biol 15:665–676 [157]. The lumen can appear through four different morphological mechanisms. In the case of tubular cavitation, lumen is formed at the contact surface between the cells; in the case of cellular cavitation, vesicles fuse into the cytoplasm, forming a large intracellular lumen that expands and fuses to the plasma membrane of the cell thereafter; invagination of the plasma membrane creates a lumen by adding and incorporating portions of the membrane into the cell; in the case of multicellular cavitation, a lumen is formed when the cells in the center are removed by apoptosis and the cells of the periphery are polarized. Tubular cavitation and invagination appear to be the predominant mechanisms in the case of blood vessels

elongation factor formin-like 3 (fmnl3) in this phenomenon [163]. The separation is enhanced by the action of VEGF, which, by activating VEGFR2, induces a change in the shape of the cell. After formation of the initial lumen, which is produced by the joining of two endothelial cells, a global remodeling takes place. To this end, a positional rearrangement of the cells takes place inside the vascular tube. Cells, instead of being joined to one another, are superimposed. Homotypic contact (contact between two places in the same cell) is established, which advances in a similar way to a zipper. This results in positional tilt and superposition of two endothelial cells.

In the case of invagination, the fusion phenomenon occurs in the vicinity of a pre-existing lumen, in a proximal-distal orientation, and depends on the blood flow. Here, the invaginating internal cell membrane undergoes a concomitant apico-basal polarization. Vascular lumen is formed by the extension of transcellular light through the endothelial cell. This results in the formation of a hollow unicellular tube. These two phenomena, tubular cavitation and invagination, can take place on the same vascular tree.

However, it should be noted that these processes were clearly identified in the zebrafish embryo. It is very probable that these same mechanisms are universal and also occur in mammals, reflected in unpublished results from different laboratories.

Recently the Gerhardt laboratory has published interesting results, providing an explanation of how hemodynamic forces contribute to lumen formation [164]. They show that blood flow drives lumen expansion in angiogenic sprouts by a process called inversed blebbing. Inversed blebbing involves the induction of spherical deformations of the apical membrane of endothelial cells.

For larger vessels, lumen must increase in size during embryonic development, which is produced by endothelial cell proliferation. This phenomenon is the basis of arteriogenesis, which means the formation of larger vessels from arterioles, plays an important role in the formation of the collateral circulation, and is stimulated by blood flow in adults. In the embryo, the stimulating factors are not well-known.

10.3 Formation of Lymphatic and Venous Valves

The valves of the lymphatic vessels and the veins have an important function because they prevent reflux of the lymphatic fluid or venous blood. The formation of the valves is initiated when endothelial cells change position and protrude into the vascular lumen.A number of factors have been identified to be implicated in valve formation. These factors include transcription factors [Nuclear Factor of Activated T-Cells (NFATc), Forkhead Box Protein C2 (Foxc2)], adhesion molecules (Integrin alpha 9, Connexins 37 and 43), and components of the extracellular matrix (fibronectin-EIIIA) [165]. More recently, the spatio-temporal development of valve formation has been described [166]. This takes place in four stages: initiation, condensation, elongation, and maturation. As with other processes described in developmental biology, initiation of valve formation begins with a selection of cells to initiate the valve forming process. Cells having the ability to form valves express high levels of the Prospero Homeobox-1 (Prox1) transcription factor, the main gene of lymphatic determinism, and of Foxc2. Prox1 and Foxc2 positive cells are surrounded by Prox1 and Foxc2 negative cells. There is therefore a limitation of the area able to form valves, and it appears that intracellular factors such as calcineurins/NFAT are involved. During the condensation phase, cells adopt a cuboid shape and close contacts are established, which require the presence of functional proteins (involved in adhesion between two cells) called Connexins 37 (Cx37). The proposed model is that calcineurin/NFAT is required to induce overexpression of Prox1 and Foxc2 in selected areas, which, in turn, activates the expression of Cx37. The elongation phase is accompanied by production of proteins and matrix adhesion receptors (Integrin α9) expressed at the cell surface. During the maturation phase, a thick extracellular matrix is deposited. The repertoire of valve-forming factors continues to grow. For example, recently Bone Morphogenetic Protein-9 (BMP9) and Elastin Microfibril Interfacer-1 (EMILIN1) have been described to be involved in the formation of the valves [165]. However, the precise underlying mechanisms are not yet clear.

10.4 Vascular Permeability

The contact zones between endothelial cells have a very important role to ensure vessel stability. These areas contain transmembrane proteins which establish contacts allowing cohesion between two endothelial cells. In addition, they are involved in the control of vascular permeability to fluids and molecules, and allow transmigration of white blood cells. The junctional components are also involved in sprouting because breaking of cell junctions allows the dissociation of cells to free them from junctional constraints and their subsequent migration out of the preexisting vessel.

Occludin and VE-cadherin are major junctional molecules [167]. They are part of the tight and adherent junctions, respectively. Occludin is a protein with four transmembrane domains expressed, for example, in the endothelium of poorly permeable vessels such as the blood-brain barrier. VE-cadherin comprises five immunoglobulin-like domains in the extracellular region, a transmembrane domain and a small intracellular sequence. The extracellular region of VE-cadherin establishes so-called "homophilic" contacts between two molecules expressed at two different cell types. To be fully functional, VE-cadherin molecules must be grouped on the cell surface, and it is the transmembrane domain that plays an important role here. The cytoplasmic part of VE-cadherin binds to intracellular proteins called catenins, which serve to stabilize intercellular adhesion. In addition, VE-cadherin is also linked to the actin cytoskeleton via the catenins.

Angiogenic factors such as VEGF and Angiopoietin-1 may modify these intercellular adhesions to allow vascular permeability and initiation of the vascular sprout [168] (Fig. 10.4). On the other hand, Fibroblast Growth Factors (FGF) and Sphingosine-1 phosphate (a lipid-like molecule) stabilize the adhesions. These molecules seem necessary to maintain cohesion of the vascular tube (see Section 10.5.4 "The Return of FGFs").

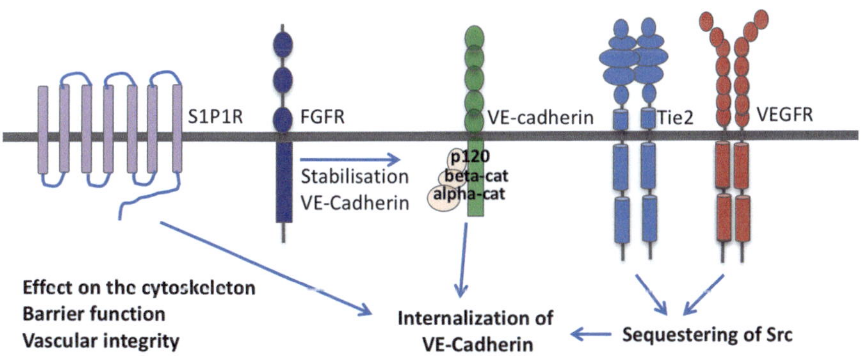

Fig. 10.4 Effects of angiogenic factors on intercellular adhesions. Overall, there are two types of factors: so-called destabilizing factors (VEGF, Angiopoietin-1) and stabilizing factors (FGF, S1P). The destabilizing factors induce the internalization of the VE-cadherin at the cell surface and thus decrease the content of this molecule on the cell surface. The effects of FGF and S1P are opposite. Figure redrawn and modified by the author of Azzi S, Hebda JK, Gavard J, 2013 Frontiers in Oncology, Vol 3, Sect 211 [169]

10.5 Other Cellular Factors and Interactions

Is the sole existence of VEGF and in particular of the classical form of this factor (VEGF-A) sufficient to explain angiogenesis? The answer is "no" because, in some tissues and tumors, angiogenesis is clearly independent of VEGF. On the other hand, other cellular interactions appear to be very important for vascular development, such as the interaction between pericytes and endothelial cells or the modulating role of blood cells on the vasculature. What are these other types of molecules and cellular interactions?

10.5.1 VEGF-C

VEGF-C is a molecule of vital importance in lymphangiogenesis as we see later in more detail [170]. Nevertheless, VEGF-C can also stimulate the growth of blood vessels. How can we explain this, because VEGF-C interacts with the VEGFR3 receptors and they are expressed in particular on lymphatic vessels?

VEGF-C is produced in the form of precursors and matured by enzymes called convertases [171]. These convertases cleave the "precursor" form from 419 amino acids (59 kDa) to a 31-kDa form that is converted into a 21-kDa form. This maturation has consequences concerning interaction with VEGF receptors. VEGF-C (31 kDa) has a very strong affinity for VEGFR3 and this is important for the regulation of lymphangiogenesis. On the other hand, the 21-kDa form seems to interact with VEGFR3 and VEGFR2 and therefore has an effect on both angiogenesis and lymphangiogenesis. Moreover, as stated in one of the previous chapters, Ralf Adams's team [149] (Max Planck Institute, Münster, Germany) reported that VEGFR3 is also expressed in endothelial cells at the level of tip cells. However, the importance of the presence of VEGFR3 for the angiogenic function is debatable. More recently, the team of Kari Alitalo, a professor at the University of Helsinki, brought new elements in this debate as already briefly mentioned [150]. Using conditional knock-out mouse models, they showed that the absence of VEGFR2, but not of VEGFR3, impaired sprouting of retinal vessels. Furthermore, expression of VEGFR3 could not compensate for the loss of VEGFR2. The absence of VEGFR3 led to an increase in the sprouting of retinal vessels. In addition, Kari Alitalo's team described a feedback interaction between VEGFR2 and VEGFR3. Indeed, VEGFR3 deletion induces the increase in the expression of VEGFR2 which, on the other hand, stimulates the expression of VEGFR3, which ultimately decreases VEGFR2. In this concept, VEGFR3 would rather function as a negative regulator of VEGFR2, which keeps the central role in angiogenesis. It should be emphasized that these conclusions are derived from the use of a single experimental model, the retinal vascularization model. It is not certain that these conclusions can be generalized and that VEGFR3 has no significant function in the blood vasculature outside of its modulatory role on VEGFR2.

10.5.2 PLGF

Placental Growth Factors (PLGF), isolated, as the name suggests, first from the placenta, have the peculiarity of not binding to VEGFR2 but only to VEGFR1. As already mentioned, VEGFR2 is the primary receptor by which VEGF regulates angiogenesis. The team of Peter Carmeliet showed that the absence of the PLGF did not lead to embryonic abnormalities in mice [172, 173]. On the other hand, in adults, angiogenesis and tumor growth were disrupted. PLGF seems to regulate different stages of the vasoformation, including endothelial cell migration and sprouting. Furthermore, because PLGF interacts with VEGFR1 it also stimulates the recruitment of inflammatory cells (monocytes, etc.) which migrate to angiogenesis sites. These cells then produce factors that, in turn, stimulate angiogenesis. However, Yihai Cao's team (Karolinska Institute, Stockholm) reported somewhat different results [174, 175]. According to his team, PLGF induces the formation of dilated vessels, and loss of PLGF in tumor cells would promote tumor growth. If for Peter Carmeliet and Yihai Cao the PLGF has pro-angiogenic effects, the meaning of this effect is totally different. We see later the implication of these observations for the therapeutic inhibition of angiogenesis and, in particular, of tumor angiogenesis.

10.5.3 Angiopoietins

Another family of molecules, known as angiopoietins, was discovered and characterized for its effects on vascular development [176]. Two forms have been extensively studied: Angiopoietin-1 and Angiopoietin-2. It should be noted that the main role of these factors is to modulate the sensitivity of the blood vessel to other angiogenic factors and mainly to VEGF. They are not angiogenic stimulators in the strict sense but factors modifying the quality of vessel (Fig. 10.5). During development, these factors contribute to vessel maturation, which is an important step in the stabilization of the blood vessels by tight attachment of pericytes to the vascular tube. In this regard, angiopoietin-1 has opposing effects to angiopoietin-2. Angiopoietin-1 promotes the interaction between pericytes and the vascular tube, whereas angiopoietin-2 destabilizes this interaction to make the vessel rather immature. An immature vessel can be stimulated by VEGF, whereas a mature vessel is refractory to this stimulation. This means that angiopoietin-2 stimulates angiogenesis and that angiopoietin-1 is a quiescence factor. Nevertheless, there are exceptions to this rule, and angiopoietin-1 may also, in some cases, directly affect sprouting of blood vessels. In addition to developmental angiogenesis, angiopoietin-1 and angiopoietin-2 also play a role in pathological angiogenesis, as seen in cancer or in inflammatory diseases and tissue repair [177].

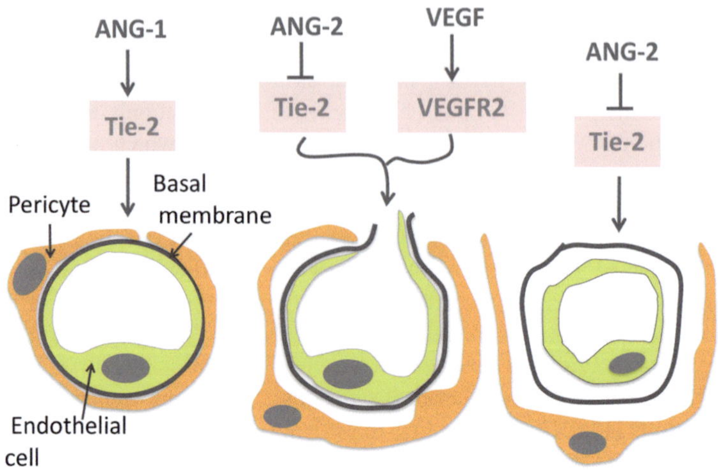

Fig. 10.5 Concept of vessel maturation and role of angiopoietins. Angiopoietin-1 (Ang-1) leads to the narrow attachment of pericytes to the basement membrane and endothelial cells composing the vascular tube by binding to the Tie-2 receptor. In the case of a tumor, Angiopoietin-2 (Ang-2) is produced in excess and counteracts the effect of Ang-1 by binding competitively on the same receptor (Tie-2). VEGF can then stimulate endothelial cells and lead to growth of the vascular tube. Excess Ang-2 in the absence of VEGF stimulation may lead to regression of the vascular tube. Figure of the author

10.5.4 The Return of the FGFs

With the advent of VEGF, FGF was no longer considered to have a role to play in angiogenesis. This conviction was reinforced by the fact that gene inactivation of some of the FGF family members (FGF-1 and FGF-2, for example) did not show any obvious vascular phenotype. It was then thought that FGF was dispensable to angiogenesis. Nevertheless, it should be noted that FGF has 23 members and 4 main types of receptors identified. This can lead to functional redundancy between these different molecules. This means that the absence of one of these molecules could be compensated by another member of the family.

Furthermore, several observations have been made subsequently that support a role for FGF in angiogenesis. First, FGF2 can induce autocrine or paracrine VEGF-A loops. This has been observed in endothelial cells in culture but also in tumor cells such as cells derived from brain tumors [178]. Furthermore, it has been shown that the blocking of VEGF-C inhibited FGF2-induced lymphangiogenesis [179]. This indicates that FGF could have an indirect role in both angiogenesis and lymphangiogenesis by stimulating the expression of VEGF-A and VEGF-C. Apart from the indirect effect of FGF on lymphangiogenesis, a direct effect on the lymphatic endothelium has also been described [180]. This is described in more detail in the chapter “Lymphangiogenesis”. Second, my laboratory has shown that the functional blocking of FGF in the retinal pigment epithelium (EPR) results in a considerable decrease in the choroidal vessels and in the absence of formation of the retinal vessels [181]. This indicates that the assembly of the endothelial cells is defective when the function of the FGFs is inhibited. Third, in preclinical models using RipTag mice treated with VEGF inhibitors, tumor escape was observed. The escape has been shown to be caused by the induction of FGF expression [182]. Fourth, Michael Simons' team at Yale University has shown that FGF is important in maintaining vascular integrity [183], which corroborates our previous studies. Indeed, it seems that the connection between endothelial cells is controlled by FGF. The inhibition of these molecules causes a disorganization of vessels and a rupture of the tubular structure, which is in agreement with our observations on the retina. This effect is produced by the interference of FGF with the function of the cadherin adhesion molecules, which are essential for maintaining the contact between endothelial cells, thus contributing to the maintenance and stabilization of the vascular tube. These observations were recently reinforced using the zebrafish model and a chemical inhibitor of FGF, the SSR128129E [184]. This molecule is an allosteric inhibitor of FGF receptors. The inhibitor affects vessels sprouting but has, more importantly, a drastic effect on vascular integrity in this model by disrupting the cellular junctions between endothelial cells dependent on VE-cadherin. This is, therefore, a very important argument, demonstrating that developmental angiogenesis also requires the presence of FGFs to maintain cohesion of the vascular tubes.

Most importantly, the Simons laboratory has recently identified a role for FGFs in the metabolism of vascular cells in both blood and lymphatic vessels [185]. They have shown that FGFs control c-MYC (MYC) expression, which in turn induces the glycolytic enzyme hexokinase-2 (HK2). A defect in FGF signaling leads to a decrease in HK2, which impairs EC glycolysis. This impacts on migration and proliferation of ECs. HK2 overexpression can partially compensate for the defect in FGF signaling. Thus, FGF-dependent regulation of endothelial glycolysis is a critical process in developmental and adult vascular growth and development.

10.5.5 Role of Pericytes and Platelet-Derived Growth Factors (PDGF)

Apart from endothelial cells, another cell type called pericytes is important in the constitution of the vascular tube. These cells cover the periphery of the blood vessel and are closely associated in a normal vessel with endothelial cells. Christer Betsholtz (now University of Uppsala) has shown that these cells are involved in the stabilization of the vascular tube and elucidated, at least partially, the mechanism of pericyte coverage [186]. He has shown that growth factors called Platelet-derived Growth Factors (PDGF) are involved [187] and that a defect of these molecules or their receptor causes vessel destabilization. Vessel destabilization causes the passage of blood into the tissues, easily passing through the vascular wall. This has important consequences for retinal or choroid vessels in the eye. During diabetic retinopathy, vessel destabilization occurs. which leads to leakage of liquids to the tissues ("exudates") and to vascular cell proliferation. This causes damage to the retina, leading to vision loss. Choroidal vessels which are located in the subretinal space may also be affected by this destabilization. Pericytic coverage also appears to be

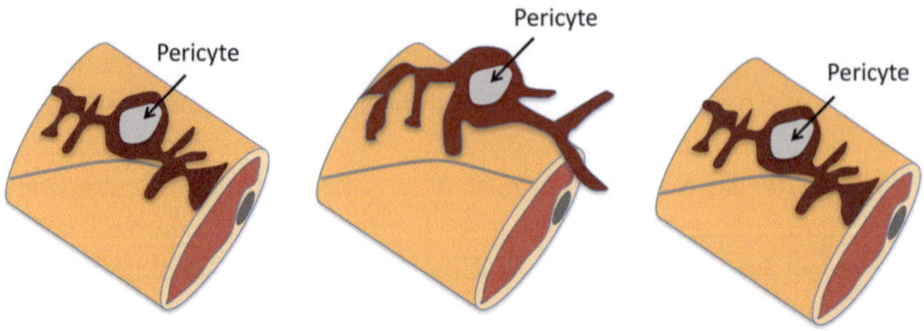

Fig. 10.6 Pericyte-endothelial cell relationships and stabilization of blood vessel. This figure illustrates the interaction between endothelial cells and pericytes that, as we have seen, depends on Ang1/2 and PDGFs. Normal vessel (*top*). Tumor vessel (*middle*). Tumor vessel after treatment with a VEGF inhibitor (*bottom*). It is seen that the tumor vessel has a loose pericytic coverage, which after anti-VEGF treatment is normalized. Figure of the author

important for "vascular quiescence," as already indicated [188]. The term "vascular quiescence" is understood as a state where blood vessels no longer respond to stimulation by the angiogenic factors or inhibitors. Vessels with good pericyte coverage (close association of pericytes to the vascular tube) are no longer likely to respond to stimulation or inhibition by external factors (Fig. 10.6). This should make normal (non-pathological) vessels theoretically refractory to angiogenesis inhibitors and, thus, should limit the extent of side effects associated with anti-angiogenic therapy. However, this is less true than originally assumed. Indeed, it has been observed that treatment of patients with Avastin, a potent inhibitor of VEGF-A, may result in changes in normal non-tumor vessels. This indicates that normal vessels are less stabilized than originally thought.

As far as inflammatory cells are concerned, their role is beginning to be better understood. Different types of inflammatory cells can be distinguished, including neutrophils, monocytes, and lymphocytes. All these cells can modulate angiogenesis. The team of Napoleone Ferrara, when still working at Genentech, showed that a certain class of white blood cells (CD11b+Gr1+ neutrophils) produces additional angiogenic factors [189]. This may even be the reason for resistance to certain anti-angiogenic therapies. In addition, the same team has also shown that IL17, an inflammatory cytokine produced by lymphocyte cells (called TH17 cells), stimulated the number and function of these CD11b+Gr1+ cells. Thus, activation of TH17 cells may contribute to resistance to anti-angiogenic therapies [190]. Monocytes and macrophages also have a role to play in the angiogenic process. It is generally accepted that macrophages are derived from monocytes. There are generally two large groups of macrophages called M1 and M2 macrophages. M1 macrophages are classically stimulated by interferon and lipopolysaccharide and have a role in acute inflammation and cytotoxicity. The team of Richard Lang (University of Cincinnati, USA) showed that macrophages have an active role in vascular regression by activating a programmed death pathway during developmental angiogenesis using "Wingless" signaling (WNT) [191]. This gene has been identified as an important regulator of morphogenesis in *Drosophila*. The WNT pathway is indeed an important signaling pathway for morphogenesis during vertebrates and invertebrates development and plays an important role in angiogenesis [192]. M2 macrophages, on the other hand (which can be subdivided into M2a, b, etc.), are conventionally stimulated by interleukin-4 (IL4) and are involved in chronic inflammation, repair, and tissue remodeling. These macrophages can accumulate around vessels in situations of ischemia or in tumors, where they exert a stimulating role of angiogenesis. The role of macrophages has been elucidated in an elegant manner by Jeffrey Pollard (Albert Einstein College of Medicine, New York), who has shown that the absence of macrophages results in significant defects in vascularization [193]. It has been shown that monocytes and M2 macrophages produce FGF2, chemokines, and PLGF, the primary stimulatory role being attributed to FGF2 [194]. This indicates a paracrine action of the M2 macrophages on vascular development. It is thus clear that macrophages have diametrically opposite effects as a function of their polarization in macrophages M1 or M2. It is commonly accepted that macrophages are derived from monocytes as indicated above. However, the recent work of Fréderic Geissmann (Kings College, London) has led to a totally different view of the macrophage ontogenesis. Indeed, using a genetic approach in mice, he showed that the resident macrophages of the tissues did not originate from the blood hematopoietic cells and were therefore not derived from the circulating monocytes, but were derived from the allantois [195]. Moreover, the subdivision between macrophages M1 and M2 is artificial, according to Frederic Geissmann, and does not, therefore, correspond to a biological reality. This vision is not (yet) accepted fully by the scientific community and has caused controversy. It can be said that this is, in a way, an updating of the controversy between the blastema and corpuscular theory in a modern form!

10.5.6 Microglial Cells

If microglial cells have many similarities with macrophages, their origin is quite different. Microglial cells are derived from the yolk sac and migrate during embryogenesis in many tissues, especially to the nervous tissue. Microglial cells have close interactions with blood vessels as well as with other tissue components, such as astrocytes or neurons. Their role in angiogenesis is less well-known. Using transgenic mouse models with genetic invalidation of microglial cells, retinal vessels were found rarefied [196]. Microglia cells are usually associated with tip cells at locations where two tip cells meet. It appears that microglial cells secrete a factor distinct from VEGF which exerts a stimulatory role on tip cells [196]. The nature of this factor is not known at this time. An American-Chinese team described a factor called p53-Upregulated Modulator of Apoptosis (PUMA), whose invalidation leads to a reduction in microglial cells and a decrease in angiogenesis [197]. The factor secreted by the microglial cells could be identical to PUMA.

10.6 Angiogenic Switch

Angiogenesis is preceded by molecular events that are responsible for its activation. These events are under the control of a mechanism known as the angiogenic switch. It occurs in all tissues that emit an angiogenic signal, such as cancerous, inflammatory, or ischemic tissues.This "switch" was experimentally demonstrated by Douglas Hanahan (University of California, San Francisco and ISREC, Lausanne) and Judah Folkman [198]. Douglas Hanahan had developed a genetic model in mice, called the RipTag mouse model (for details see later in the book). This mouse develops an insulinoma that originates in the β cells of the pancreas. This tumor progressively develops in the pancreas and passes through different stages. First, hyperplastic lesions appear which are not vascularized. At a later stage of development, these tumors vascularize and then evolve into invasive tumors. There is, therefore, the initiation of angiogenesis at a given stage of tumor development. This corresponds to the initiation of a signaling cascade leading to the production of angiogenic factors.

What is the nature of this switch? A first group of molecules has been identified which are called "Hypoxia-inducible Factors" (HIFs) by Greg Semenza's team (Johns Hopkins University, Baltimore) [199]. These factors are transcription factors, which means that they can regulate the expression at the genetic level of other molecules responsible for the given biological effect. HIF molecules bind to nucleotide sequences called the hypoxia response element or HRE located in the promoter region in target genes. Thus, the VEGF gene harbors in its promoter an HRE sequence and, thus, transcription can be stimulated if the transcription factor binds to these HRE elements. How is this then achieved? HIF must be stabilized in the cytoplasm because, as with any other protein, it has limited survival and is degraded in the cell after a certain time. Stabilization is carried out by a biochemical mechanism called hydroxylation. A chemical group, a hydroxyl (-OH), is incorporated at the level of the HIF protein and this promotes its degradation. On the other hand, in its non-hydroxylated form, HIF penetrates into the cell nucleus and binds to the HRE elements, together with cofactors, to induce the expression of target genes. In its hydroxylated form, HIF binds to other proteins where it is transferred to the proteasome. The proteasome is a molecular complex that then degrades HIF. Hydroxylation is carried out by enzymes called Prolyl Hydroxylase (PHD) which, precisely, have the ability to incorporate an -OH group in the HIF protein [200]. A schematic representation of the regulation of the HIF system is given in Fig. 10.7.

Apart from this main system of regulation, other mechanisms exist. The team of Jacques Pouyssegur of the Institute of Aging and Cancer of Nice showed that the MAP-kinase pathway also contributed to the amplification of the signal induced by HIF [201]. The MAP kinase pathway is one of the main pathways that propagates the signal induced by many growth factor or cytokine receptors from the plasma membrane inside to the cell.Other regulators of the angiogenic "switch" are represented by genes directly involved in cell transformation, such as the oncogenes. Oncogenes were discovered more than 20 years ago and are important in the transformation of tumor cells. Among the oncogenes, the angiogenesis role of two of them has been extensively studied, including the c-myc and c-ras oncogene. It has

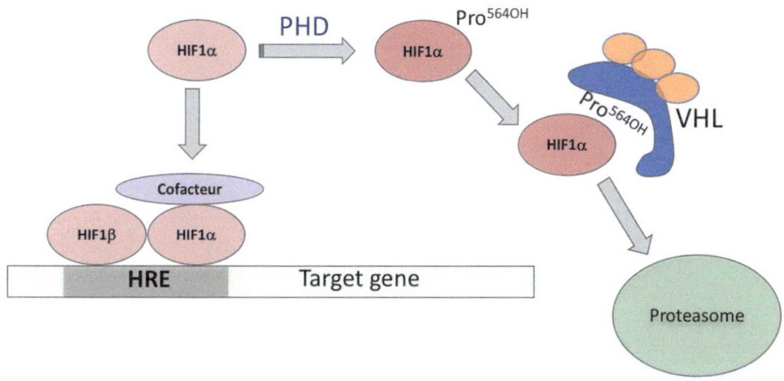

Fig. 10.7 Regulating VEGF expression using HIF and Prolyl Hydroxylase (PHD). In its non-hydroxylated form HIF1α acts as a transcription factor by binding to the HRE motif ("Hypoxia response element") of the target genes. In its hydroxylated form (Pro5460H), it is routed to the proteasome to be degraded. A number of molecular partners also intervene such as HIF1β and the product of the von Hippel-Lindau (VHL) gene which associates with HIF1α hydroxylated to direct it toward the proteasome. Drawing by the author

been shown that the c-myc oncogene can regulate VEGF, thrombospondin-1, and other angiogenesis factors [202]. Furthermore, inflammatory cells can also contribute to the angiogenic "switch."

10.7 Signaling Induced by Angiogenic Factors

As with any cellular stimulating factor, angiogenic factors must interact with receptors located on the surface of vascular cells. As already mentioned above, receptors for VEGF have been identified named VEGFR1, VEGFR2, and VEGFR3, as well as two co-receptors named Neuropilin-1 (NRP1) and Neuropilin-2 (NRP2) [191]. VEGFR1-R3 are tyrosine kinase receptors because the tyrosine kinase domains are essential for enabling signal transmission within the cell (Fig. 10.8).

Tyrosine kinase domains may phosphorylate themselves by a mechanism called autophosphorylation. During autophosphorylation, two VEGF receptor monomers associate with each and forms a dimer. Autophosphorylation is initiated by the close association of the two monomeric units after ligand binding. To phosphorylate the receptor completely, a mechanism called a "ping-pong" has been described (Fig. 10.9). This mechanism is characterized by alternating kinase activity and substrate between the two monomeric units, leading to the complete phosphorylation of the receptor dimer.

Following autophosphorylation, molecules located inside the cell (signaling modules, adapters, etc.) bind to the phosphorylated segments of the receptor to initiate a signaling transduction cascade. There are different pathways that have been described such as the MAP-kinase pathway, which essentially regulates

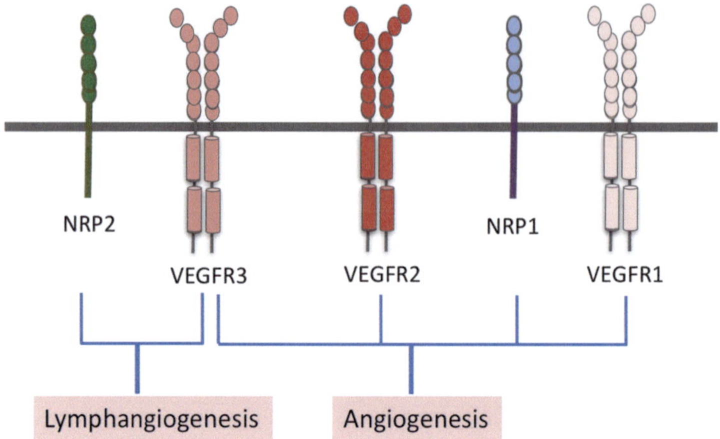

Fig. 10.8 Different forms of VEGF and related molecules and their interaction with their receptors. As shown in the figure, VEGF-A binds to VEGFR2 and VEGFR1, with Neuropilin (NRP1) being a co-receptor in these interactions. VEGF-C and VEGF-D both bind VEGFR3 and VEGFR2. In this case, NRP2 is the co-receptor. VEGF-C and VEGF-D primarily regulate lymphangiogenesis. VEGF-B and Placental growth factor (PLGF, a molecule related to VEGF) only interact with VEGFR1. Figure of the author

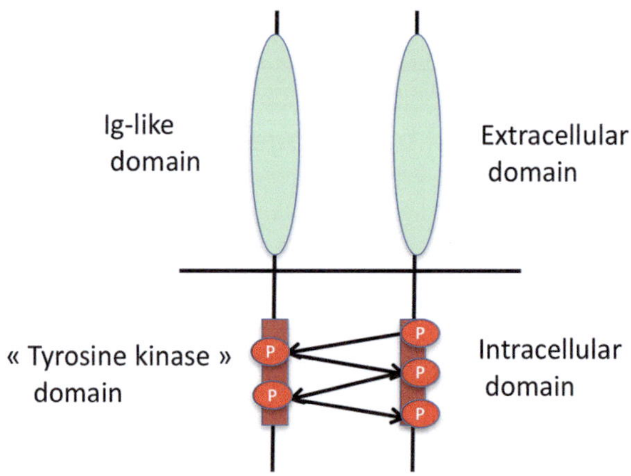

Fig. 10.9 "Ping-pong" mechanism of receptor autophosphorylation

proliferation and cell migration or the PI3 kinase pathway, which controls cell survival [203] (Fig. 10.10).

Apart from the main receptors (VEGFR1, VEGFR2, VEGFR3), co-receptors have also been described (Fig. 10.8). For VEGF, two co-receptors were identified, Neuropilin-1 (NRP1) and Neuropilin-2 (NRP2), which we have already encountered in previous chapters. NRPs play an important role in the nervous system where they control axonal guidance (see later in the book). In the vascular system, their function

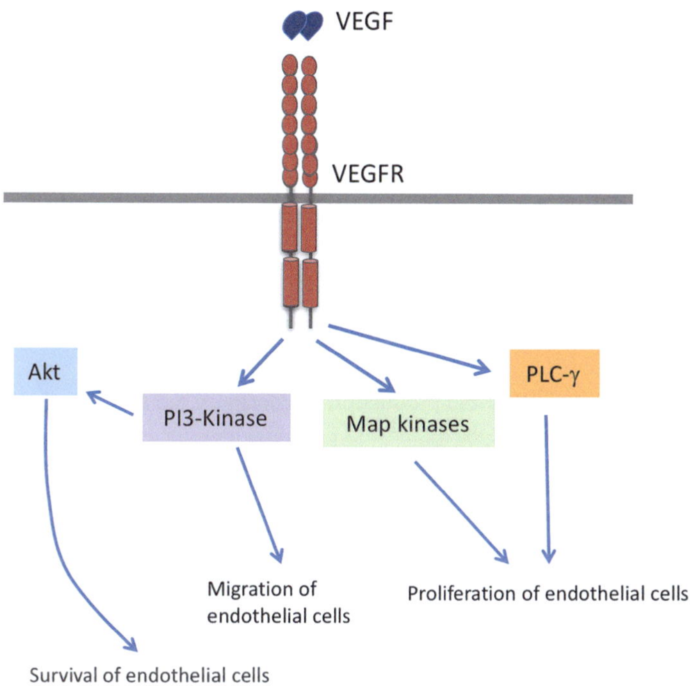

Fig. 10.10 Signaling pathways induced by VEGF. VEGF receptors dimerize and autophosphorylate. This causes the initiation of signal transduction pathways that propagate the induced signal to the interior of the endothelial cell. Three pathways are initiated including the MAP kinase pathway, the PLCγ pathway, and the so-called PI3 kinase pathway. If these pathways have different functions, they are interdependent and sometimes even redundant. Figure by the author

is to enhance the interaction of VEGF with the main VEGF receptors. NRP1 is mainly expressed in the arterial system and NRP2 in the venous and lymphatic system. This indicates different roles of these co-receptors, depending on their location on the vascular tree [204].

Many other regulators, besides VEGF, induce signaling in the vascular cells. These include FGFs, which activate tyrosine kinase domain receptors, chemokines, which activate seven transmembrane domains receptors, Transforming growth factor (TGF-β), which activates serine/threonine kinase domain receptors, and netrins, which activate dependence receptors. The signaling mechanisms for these receptors are not described in detail here. For more detailed information, the reader may refer to a series of excellent reviews on this subject [205–207].

Another aspect of the angiogenic signaling is represented by what may be called "indirect" signaling. In this case, a cell may be activated by a factor which is not per se an angiogenic factor. The cell may produce an angiogenic factor, which then is able to stimulate vascular cells. This is, for instance, the case for inflammatory cells. Monocytes, for example, differentiate in tissues in macrophage types M1 or M2 as indicated previously. The cytokine-stimulated M2 macrophage has pro-angiogenic characteristics and can produce factors such as VEGF and thus contribute to the angiogenic "switch."

The endothelium is the main place in which these mechanisms come into play because it is the central actor of angiogenesis. However, for stabilization and vascular maturation, signaling at the level of the pericytes and the smooth muscle cells is required. This has already been discussed above and involves PDGF signaling and other factors that are implicated in the recruitment of pericytes and smooth muscle cells to the vessel.

10.8 Metabolism and Angiogenesis

Metabolism is important for the production of energy in the cell. This results in the production of Adenosine triphosphate (ATP) which then feeds the various biochemical reactions in the cell. There are two major ATP production sites – on the one hand, glycolysis and, on the other, the citrate cycle, also known as the Krebs cycle according to its discoverer. In the latter, oxidative metabolism takes place because oxygen is required in this case. It all starts with the transport of glucose into the cell, which is then transformed via a chain of biochemical reactions into pyruvate. The pyruvate is then taken up in the Krebs cycle to form carbon and ATP. Glycolysis itself can also produce ATP. However, the production of ATP by oxidative metabolism is much more efficient than by glycolysis alone.

In tissues with high metabolic activity, glycolysis is favored and diverted to form lactic acid. This is called the Warburg effect according to its discoverer, Otto Warburg, who was director of the Kaiser Wilhelm's Institute of Cellular Physiology under the Third Reich. The Kaiser Wilhelm institutes were already funded well before the advent of Hitler by the Rockefeller Foundation, and this continued for some time, even after the advent of the Third Reich. Otto Warburg, despite the fact that he was half Jewish by his father, was one of the few people to be preserved from Hitler's racial policy because Hitler had a pathological fear of cancer. For his work on cellular respiration, Warburg was awarded the Nobel Prize in 1931. Warburg postulated that cancer was caused by a dysfunction of cellular respiration which resulted in the independence of the cancer cell from oxygen and in the fermentation of glucose by the production of lactate [208, 209]. Tumors literally love lactate!

If the role of anaerobic glycolysis is well-understood in the tumor cell context, what is the situation for blood vessels? This has been fairly recently studied by Peter Carmeliet's team, whom we mentioned earlier [210]. They showed that endothelial cells favor glycolysis over oxidative metabolism (Fig. 10.11). Endothelial cells also accumulate lactate and possess strong expression of an enzyme called 6-phosphofructo-2-kinase/fructose-2,6-biphosphatase 3 (PFKF3B). This enzyme is an important stimulator of glycolysis, which promotes the conversion of fructose-6-phosphate to fructose-1.6-biphosphate via the production of fructose-2,6-biphosphate, an important stimulator of phosphofructokinase. In particular, glycolysis and PFKFB3 are strongly stimulated in tip cells and particularly in regions where filopodia are formed. Moreover, the same team has shown that the oxidation of fatty

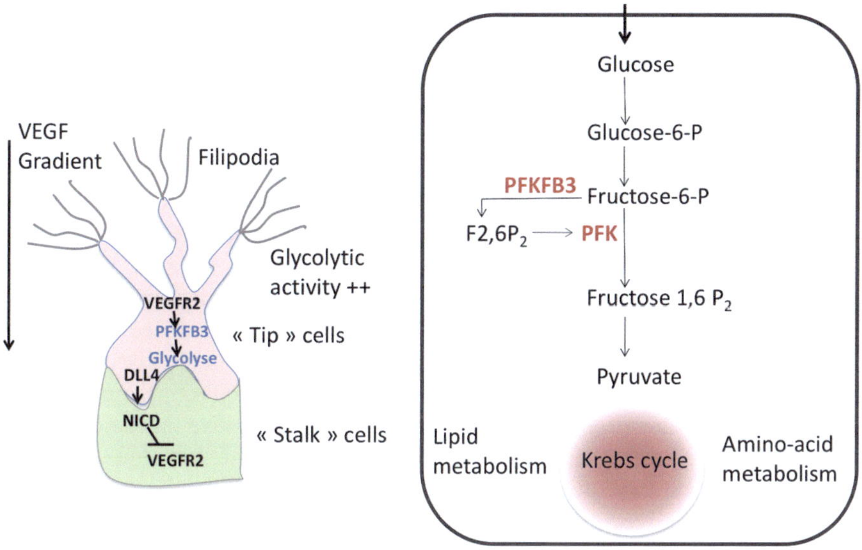

Fig. 10.11 Metabolism of endothelial cells at the tip cells. Metabolic activity is localized in tip cells with increased glycolytic activity (*left*). The pathways of energy metabolism (*right*). PFKFB3 is involved in the regulation of the transformation of fructose-6-phosphate (F6P) into fructose-1,6-biphosphate (F1,6 P2). The increase of the PFKFB3 enzyme stimulates glycolysis and promotes non-oxidative (anaerobic) metabolism. Redesigned and modified figure by the author from Verdegem D, Moens S, Stapar P, Carmeliet P Cancer and metabolism 2014, 2:19 [210]

acids is involved in the sprouting of endothelial cells [211]. Indeed, the invalidation of CPTA1 in endothelial cells, an important enzyme in fatty acids oxidation, strongly inhibits angiogenesis.

10.9 Endothelial Cell-Derived Factors Have Perfusion-Independent Effects on Organs

Another example is factors released from endothelial cells for tissue morphogenesis and organ development, representing a vessel-tissue relationship that is not dependent on the oxygenating/nourishing function of the vasculature. Indeed, inductive signals for tissue morphogenesis derived from the vasculature have been described for organ development, such as the dependency of the patterning airway branching on the proximity to the vasculature with an effect on stereotype branching, which is dramatically disturbed following vascular ablation [212]. A paradoxical relationship with non-vascular tissue exists, such as with the pancreas where, surprisingly, non-nutritional signals from blood vessels act to restrain pancreas growth [213]. VEGF-induced hypervascularization decreases pancreas size. Thus, the vasculature has a positive and negative perfusion-independent action on organ development. There is also a relationship between blood vessel development and

Langerhans Islet formation [214, 215]. VEGF seems not to be required for the development of all pancreatic islet capillaries. Furthermore, it has been shown that signals from the endothelium induce pancreatic islet formation and that, in a second step, pancreatic islet cells signal back to the endothelium via VEGF for the maintenance of the islet microcirculation.

Thus, the endothelium provides angiocrine signals that are important for an array of functions including tissue specification, patterning, organ regeneration, and maintenance of cellular functions [216]. Indeed, disruption of angiocrine signals from the endothelium impairs stem cell function and organ regeneration. In the brain, these angiocrine signals target neural stem cells and include Neurotrophin-3 (NT-3), Ephrin-B2, and Jagged-1, which maintain them in an undifferentiated state. Other angiocrine factors exhibit stimulatory activities and include BDNF, PEDF, Betacellulin, Placental Growth Factor-2 (PlGF-2) and VEGF-C, which activates NSCs into transit amplifying cells and neuroblasts. Regenerative processes involving angiocrine signaling are also dependent on other stem cell types, such as hematopoietic or spermatogonial stem cells.

Endothelium-derived signaling not only participates in developmental processes but also has a role in pathology. An example is morphogenic factors derived from the endothelium which directly impact disease, such as hypertension of the pulmonary artery (PAHT) where deficiency of apelin in pulmonary endothelial cells induces an increase in FGF2 via miR-424 and miR-503 release to stimulate smooth muscle cell multiplication and vessel thickness [217]. Elabela/toddler, which functions as a natural PAPJ receptor agonist, has been shown to be down-regulated in PAHT [218]. Elabela is detected in endothelial cells and may play a similar role to apelin in the EC–SMC interaction. Other signaling molecules for idiopathic PAHT have been described, including the Wnt/planar polarity cell pathway [219]. In addition, it has been shown that endothelial cells are able to signal to cancer stem cells in glioma and that apelin and APJ receptors are implicated (Gavard et al., personal communication). Another EC–glioma interaction has also been described involving EC-specific Pfn-1 phosphorylation, which is associated with tumor aggressiveness in human glioma [220]. Furthermore, FGF4 produced by B cell lymphoma cells activates FGFR1 and upregulates the Notch ligand Jagged-1 on neighboring ECs. EC Jagged-1 feeds back to tumor cells and induces Notch2-Hey1 [221].

These results illustrate that the nutrient-dependent function of the vasculature has been enriched with an educational role by the vasculature of the surrounding tissue in providing instructive signals.

10.10 Paradigm Revisited

The classical paradigm of J. Folkman postulated that angiogenesis was mandatory for tumors to increase beyond 2 mm in diameter. Is this statement still valid at the present time?

Two observations complicate the matter. First, Winkler and colleagues [222] performed very interesting experiments using multiphotonic intravital microscopy.

In these experiments, tumor cells are injected into the mouse and vasculature and tumor development is filmed over several days, the mice being transgenic mice expressing a fluorescent (green) protein in blood vessels and tumor cells expressing another fluorescent protein (red). These experiments have shown that angiogenesis starts much earlier than initially postulated. Only a few cells are required to stimulate vascular sprouting. Moreover, a few other cells are observed which remain immobilized in pre-existing vessels and do not induce angiogenesis. These cells resemble stem cells that initiate the vascular niche.

Moreover, there are tumor cells which present exclusively or predominantly an infiltrative behavior. These tumors use pre-existing vessels to migrate into healthy tissue, similar to railway tracks [223, 224]. This mechanism is called a vascular co-option by tumor cells. Angiogenesis, that is, the formation of new vessels, is absent in this case.

The infiltrative phenotype of tumors has been studied extensively in brain tumors [224, 225]. Indeed, glioblastoma, the most aggressive and lethal brain tumor, is characterized by the fact that it develops in an angiogenic and infiltrative manner. Initially, the tumor cells become attached to a blood vessel and develop initially without inducing new blood vessel formation. At a stage of tumor growth where hypoxia becomes more important, the tumor produces factors that then induce angiogenesis. Nevertheless, this may be variable and some cells retain their "angiogenic independence" and migrate very far into healthy tissue using different tissue structures of the brain called Scherer's structures [226]. In 1938 Scherer analyzed glioblastomas of 100 patients and observed that the tumor cells infiltrated the brain tissue by 4 different structures: pre-existing vessels, nerve fibers, interstitial tissue, and the supra-arachnoïd space. There is, thus, a close interaction between blood vessels (the angiogenic dependency of the tumor) and tumor cells that could escape this dependency and evolve in a more nutrient-restrictive environment.

The molecular mechanisms of transition between angiogenesis and tumor invasion are not fully elucidated, but four mechanisms have been described so far. The first mechanism is the expression of an activated form of the Epidermal growth factor (EGF) receptor [227]. This leads to a "mesenchymal drift" with increased tumor invasion. The second mechanism involves the Hepatocyte growth factor (HGF) receptor called c-met [228]. What is interesting in this case is that this involves the regulation of VEGF receptors in tumor cells. VEGF receptors exert an inhibitory effect on the c-met receptor. When VEGF receptors are no longer activated, the inhibitory effect is lost and c-met is activated. This leads to an increase in tumor cell invasion because c-met is a highly invasive receptor. The third mechanism involves the response to cellular stress, a mechanism described by my laboratory [229]. This involves a pathway called Unfolded protein response (UPR). The inhibition of one of the sensors of this pathway leads to reprogramming of the tumor cells, which are no longer producing angiogenic factors and thus become highly invasive. A fourth mechanisms recently described by my laboratory involves the chemokine receptor CXCR3 and Lipoprotein-related protein-1 (LRP) [230]. Indeed glioblastoma cells harbor CXCR3 and LRP1. My team has recently shown that LRP1 is a co-receptor for CXCR3 and that LRP1 increases the intracellular up-take of CXCR3 and decreases the amount of CXCR3 at the cell surface. More importantly, infiltrating tumor cells exhibit an extinction of LRP1 and increased amounts of CXCR3, thus promoting invasion.

However, these four mechanisms do not explain how tumor cells are attracted to the microenvironment and especially to vessels. It has only been shown that bradykinin produced in smooth muscle cells can attract tumor cells to the vessel [231]. Thus, much research has to be done to identify the precise mechanism of tumor cell attraction to the vessels.

Moreover, migration to vascular structures requires an adaptation of the size and shape of tumor cells. There is approximately 33% reduction in the size of tumor cells during the migration process and this reduction is caused by loss of water from the cells. Volumetric adaptation of the tumor cells is carried out using water channels, the aquaporins. These channels, however, are dependent on an efflux of Cl^- ions. In the case of gliomas, tumor cells can accumulate large amounts of Cl^- which are subsequently released by the C1 channels into the extracellular medium, followed by loss of water.

We see that the interaction between the vessels and the tumor is much more complex than initially postulated. It is not just about sprouting angiogenesis – the tumor also causes other vascular changes. Brain tumors are an example in this regard (Fig. 10.12). There is, in this case, the formation of glomeruloid structures (localized vascular growths forming veritable small "vascular tumors") and vascular loops, as well as a variety of mechanisms including sprouting, intussusception (splitting vessels), and co-option.

A totally new paradigm has been described recently by the identification of specific vessels, which are correlated with an antitumor effect. These vessels, called High Endothelial Venules (HEVs), are specialized vessels usually found in lymph nodes [232]. HEVs have been recently identified in various solid tumors and have been implicated in an anti-tumor response via the immune system [233], albeit prior reports indicate that immune cells such as NK cells are required mediators of angiogenesis inhibition by IL-12 and thus provide evidence for NK-cell cytotoxicity to endothelial cells [234]. HEVs are present in several solid tumors, such as mammary carcinoma, and trigger an antitumor immune response by allowing the influx of TH1 cells, cytotoxic effector T cells, and naïve and central memory T cells into the tumor [235]. It would be important to elucidate mechanistically how a number of these vessels could be stimulated to increase the therapeutic efficacy of immunotherapy. A step in this direction has been recently taken by the Bergers laboratory. They have shown that HEVs vessels can be stimulated by using a combination of immune-checkpoint inhibitors and anti-angiogenic therapy [236].

Another important conceptual development was the introduction of the normalization concept in tumor angiogenesis championed by the Jain laboratory [237, 238]. In this concept, anti-angiogenic treatment is destined to kill aberrant vasculature in tumors and to normalize the morphology and functionality of the

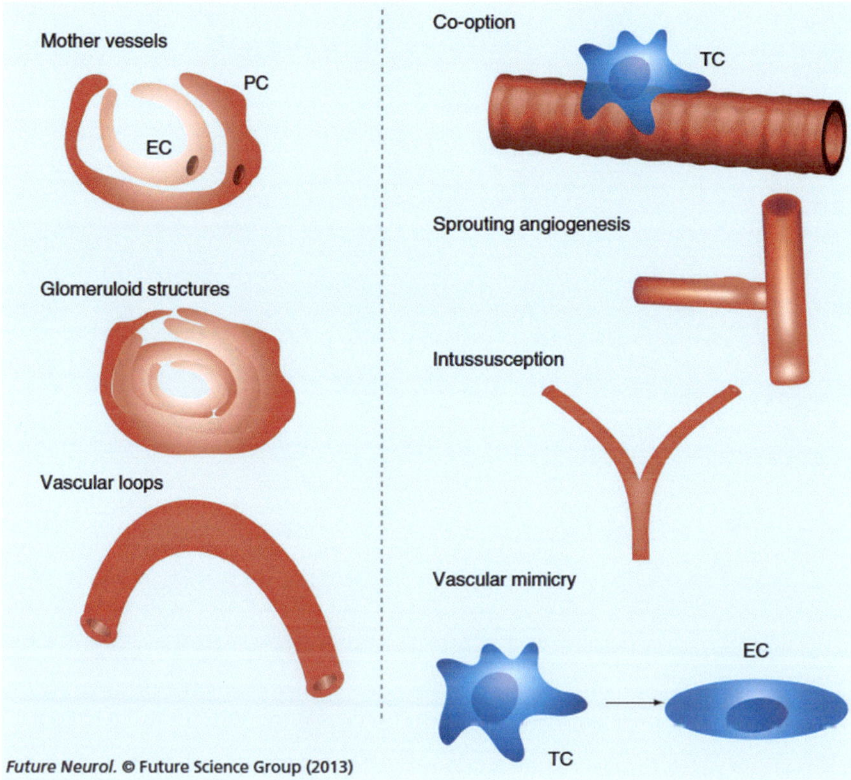

Fig. 10.12 Types of vessels found in glioblastoma and vascularization mechanisms. There is the presence of "mother" vessels that are not stabilized, glomeruloid structures, and vascular loops. The interaction with tumor cells leads to cooption, vascular sprouting, and intussuception. In addition, a mechanism called vascular mimicry has been described, implicating the transdifferentiation of tumor cells into vascular cells (see more details later in the book). Image of the author. Reprinted with permission (Javerzat S, Godard V, Bikfalvi A. Future Neurology (2013) 8(2):159–174) [239]

remaining tumor vasculature. This concept has important consequences for antitumor therapy as it permits better access of chemotherapy to the tumor because of improved functionality of the vasculature. The normalization concept has been developed for some tumor types, such as glioblastoma, but the generalization to all tumors is still a matter of debate. The normalization concept is discussed in more detail later in the book.

How to Study Angiogenesis? 11

It is important to have methods and models that allow the study of angiogenesis. Obviously, the field of angiogenesis uses all the experimental methods of cellular, molecular, and developmental biology. Nevertheless, angiogenesis research needs some very specific methods and models that require precise technical knowledge. These can be classified into three categories: in vitro models, ex vivo models, and in vivo models.

In vitro models are based mainly on the study of vascular cells and their behavior in cell culture. To study vascular cells in vitro, they must first be isolated. Endothelial cells can be isolated from multiple sources such as the umbilical cord of a newborn (endothelial cells of the umbilical vein), the large vessels (endothelial cells of the aorta, etc.), and different organs from which can be extracted microvascular cells. The process consists essentially of enzymatic digestion of tissue and isolation of endothelial cells by various methods such as attachment of cells to porous membranes or the use of gradients at different densities. Percoll, a synthetic product, has been used frequently for this purpose. Nowadays, cells are isolated using a property of vascular cells to express markers on their surfaces, which are recognized by specific antibodies. By attaching these antibodies to magnetic beads, vascular cells can be easily isolated, as these bind via the antibodies to the beads. Endothelial cells can also be isolated by flow cytometry, but the previous method has the advantage of being simpler and faster. Endothelial cells are then cultured and maintained in a suitable medium containing nutrients. We have described this and its historical developments earlier in the book.

To characterize the behavior of cells in culture, one can study cell proliferation (Fig. 11.1). For this purpose, cells are exposed to an activating or inhibiting factor and are counted after specific time intervals. This allows us to know whether a factor influences cell number, which is an indicator of an increase or inhibition of cell divisions.

In addition, cells move and have the ability to migrate in different directions. Migration of vascular cells can be studied in different ways (Fig. 11.1). A migration test can be performed by analyzing non-directional cellular motion in a horizontal

A. Bikfalvi, *A Brief History of Blood and Lymphatic Vessels*,
https://doi.org/10.1007/978-3-319-74376-9_11

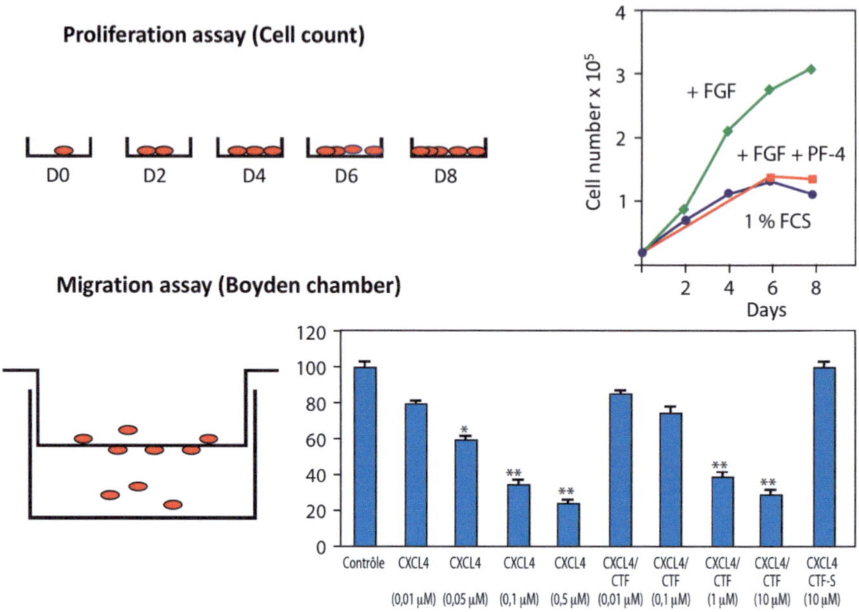

Fig. 11.1 Proliferation assay and endothelial cell migration. Proliferation assay: endothelial cells were inoculated at day 0 in the presence and absence of angiogenic factors (FGF) or inhibitor (PF4) (*top*). The cells are then counted at different time intervals. The curve on the *right* represents the influence of factors on the number of cells counted at different times. Migration assay: the cells are incubated in a Boyden chamber (*bottom*). This system comprises an upper chamber and a lower chamber, separated by a porous membrane, which can be covered with matrix proteins (fibronectin, collagen, etc.) to facilitate migration. In the upper chamber, an inhibitory agent (here CXCL4 and CXCL4CTF) is added. In the lower chamber, a growth factor (here VEGF) is added. The cells having crossed the membrane are then counted. The graph on the *right* represents the percentage of cells that migrated across the membrane in the various conditions studied. Permission to reproduce (Angiogenesis, J Robert, John Libbey Eurotext, ISBN 9782742006991)

plane. This informs us about the overall capacity to migrate in different directions. Another way to study migration is directional migration in which cells migrate as a function of a gradient. To do this, the so-called Boyden chamber, a two-compartment plastic culture dish, is used. The cells are placed in the upper compartment, and in the lower compartment the stimulator (angiogenic factor) is placed. The two compartments are separated by a porous membrane. The cells migrate through the porous membrane to join the lower compartment. The number of cells having migrated through the membrane is then counted. This, therefore, informs us about the power of the stimulating factor to induce cell migration.

A classic, very informative test is the in vitro angiogenesis assay (Fig. 11.2). This test takes into account more precisely vascular morphogenesis and the external constraints exerted by the environment. In this case, endothelial cells are cultured in a three-dimensional matrix made of collagen or fibrin gels in the presence of stimulating factors. After induction, the formation of small capillary tubes within this

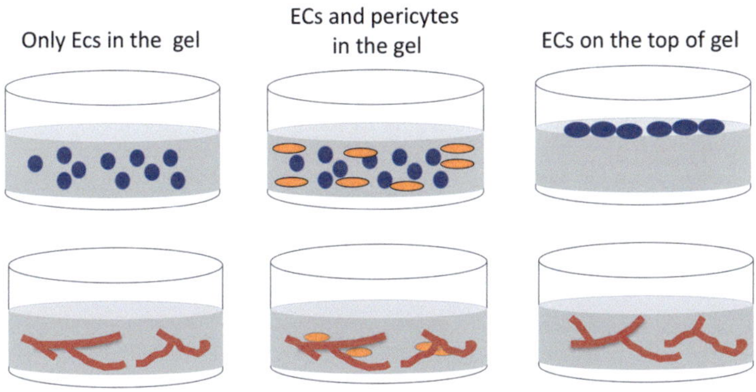

Fig. 11.2 This experiment can be done in different ways. In the first case, endothelial cells (ECs) are included in the gel. The formation of tubes is fast (about 3 days). An alternative to this model is to include endothelial cells together with pericytes. In this case, an endothelium–pericyte interaction can be observed. Another way of proceeding is to deposit endothelial cells on the top of gel and determine their ability to migrate into the gel (in the presence of factors) and to form tubular structures. Figure by the author

gel is observed. These tubes can have a lumen inside. This test exists in many variants. In general, the cells are seeded on the surface of the gel and then migrate into the interior. Another variant consists of placing cell aggregates forming spheroids within the gel. From these spheroids, angiogenic tubes form. These migrate inside the gel away from the spheroid, resembling the rays of the sun. In vitro angiogenesis has significantly contributed to characterizing the biological effects of supposedly angiogenic factors and to the study of various mechanisms such as sprouting or vascular lumen formation.

It is also possible to analyze the expression of markers on the surface of endothelial cells after stimulation with various angiogenic factors (VEGF, inflammatory factors, etc.). These analyzes can be carried out by flow cytometry using specific antibodies directed against these surface molecules and labeled with a fluorophore (fluorescent molecule) or using biochemical methods. This gives us information about the ability of factors to modulate the molecular profile of endothelial cells.

The chicken embryo model is an interesting in vivo model for studying vascular development (Fig. 11.3). The chicken embryo develops over a period of 18 days. An organ of choice for the study of vascular development is the Chorioallantoic membrane (CAM). This is an extraembryonic membrane used by the embryo to ensure gas exchange (it is the organ of respiration in the chicken embryo, so to speak). The membrane is highly vascularized and can therefore be used for many types of experiments. The team of Anne Eichmann (Yale University) showed that veins can easily be transformed into arteries by inverting blood flow [240]. This allows the expression of arterial markers in these “arterialized” veins. Our laboratory has used this model for a series of interesting experiments [230, 241–248]. Tumors can be grafted onto this membrane and functional or expression studies can be

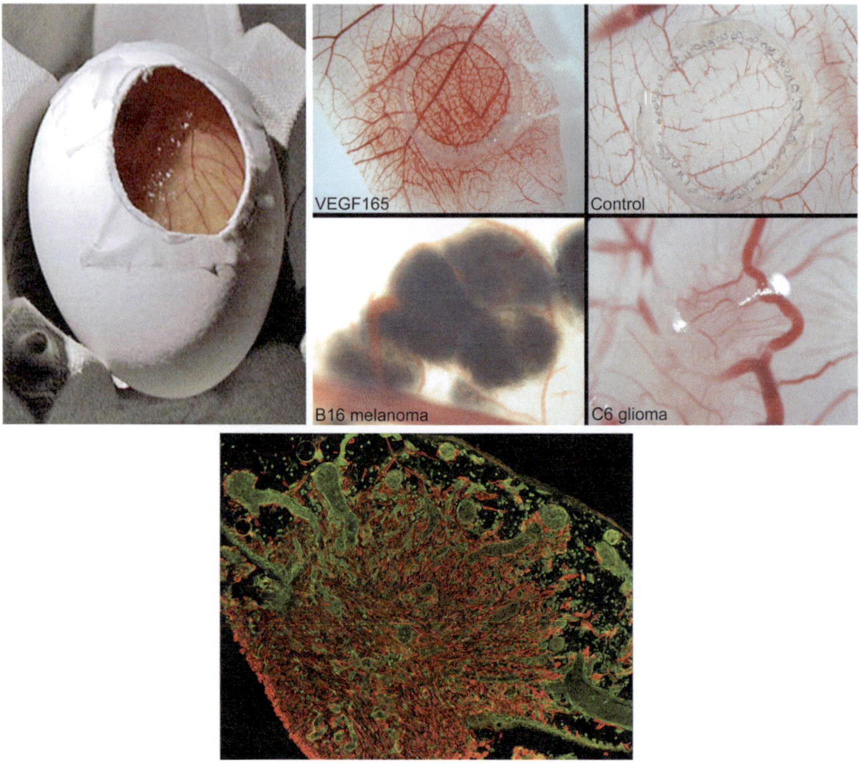

Fig. 11.3 The chicken Chorioallantoic Membrane (CAM). Part of the eggshell is removed to allow visualization of the CAM (*top left*). Four panels at the *top right*: VEGF is deposited on the CAM (*top left*), the control is on the *top right* (no VEGF deposit). We can see the very strong induction of angiogenesis on these images. Tumor cells (B16 melanoma *lower left*, C6 gliomas *lower right*) are deposited on the membrane. A strong induction of angiogenesis is seen in both cases. *Bottom* panel: Immunostaining of the histological section of a glioma implanted on the CAM. The tumor cells are represented in *red* and the blood vessels in *green*. We see that the vessels have totally infiltrated the tumor. Pictures of the author

performed. Using this model, we have shown that glioblastoma, a very aggressive tumor of the brain, can very well develop on the chorioallantoic membrane of the chicken and induce an important angiogenesis and tumor cell invasion. This is also the case for other tumors, such as renal and pancreatic cancers. An interesting and innovative use of the CAM model is to perform functional genomic studies. Indeed, one can analyze respectively the molecular changes that take place in the tumor cells and in the surrounding stroma. mRNA expression can be analyzed in this model by microarray or RNAseq sequencing. For example, if human tumor cells are implanted on the chicken CAM, a very weak overlap of the signal with the chick transcriptome is found. This means that we can obtain more or less specific information about the mRNA expression in tumor cells and in the stroma (which comes from chicken). Other methods can also be used in this model, such as proteomics (specific analysis

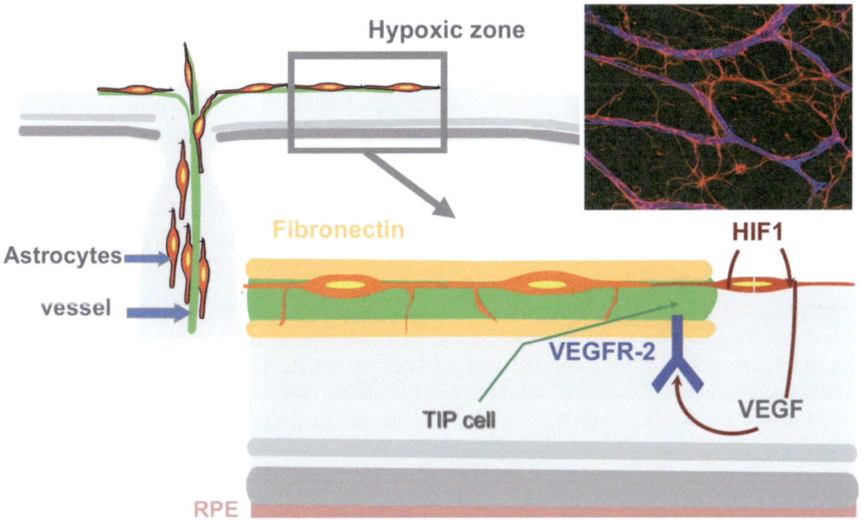

Fig. 11.4 Angiogenesis in the retina model. The central retinal vessel migrates in the optic nerve to reach the retina. The plexus is then formed by following the path astrocytes lay down. This is called "contact spanning." Astrocytes, which are at the extremity of the vessel, are hypoxic and produce VEGF. This VEGF then stimulates tip cells and this promotes angiogenesis. Figure by the author (B. Rousseau, former doctoral student)

of proteins present) and metabolomics (analysis of metabolites such as glucose, glutamine, etc.), which can provide important additional information. This model corresponds to the 3R rule of Russell and Burch (1956) for animal experiments, which should aim at improving animal welfare [249]. This rule is aimed at reducing, refining, and replacing experimental methods by methods that do not induce or that minimize suffering in animals.

The retina is another preferred model for the study of vascular development. In mice, vascularization of the retina develops up to 20 days after birth. The retinal vessels first migrate from the optic nerve to the superficial part of the retina. They follow the migration of astrocytes that secrete VEGF to promote vascularization. This interaction is known as contact spanning or extension by contact (Fig. 11.4). In a second instance, and after horizontal extension, the vessels migrate vertically to establish a second deeper plexus. We have seen in a previous chapter that "tip" cell are located at the extremity of the vascular tube. They emit filipodia which guide and orient the vascular tube. The retinal model has greatly contributed to the understanding of molecular and cellular mechanisms of blood vessels formation. For example, this model led to the establishment of the role of the Notch and Delta-like-4 genes in angiogenesis [153].

How is a functional role of a gene based on the retinal model studied? The formation of retinal vessels can be analyzed in animals that have a invalidation for a given gene. For example, animals with Delta-like-4 gene invalidation were found to have an anarchic vessel formation with the increased number of tip cells. It is also

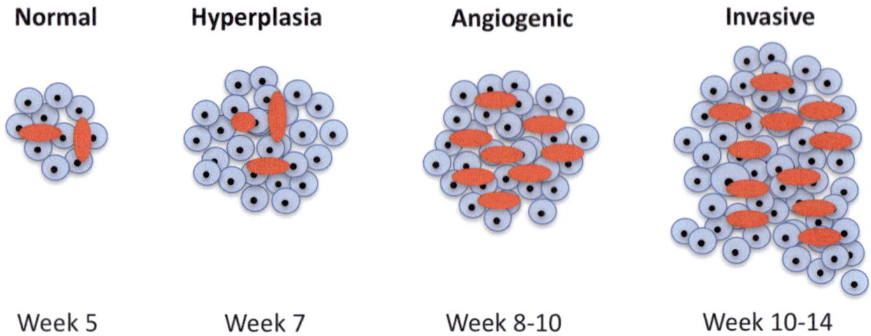

Fig. 11.5 Development of pancreatic tumor islands in the RipTag mouse model. At around 7 weeks post-natal life, pancreatic islets become hyperplastic, then vascularize at 8 weeks. Around 10–14 weeks, angiogenic islands become invasive and infiltrate the pancreatic tissue. Figure by the author

possible to inject inhibitory or activating molecules and observe the modifications induced. If, for example, a γ-secretase inhibitor (which, by the way, inhibits the Notch pathway) is administered, a very marked increase in the number of tip cells is observed.

Furthermore, the interaction between endothelial cells and pericytes can be analyzed in the retinal model. Pericytes are easy to detect because they express markers such as desmin or α2 smooth muscle actin (α2SM-actin) . It is thus possible to analyze pericytic coverage of vessels using an antibody directed against desmin or α2-SM-actin, followed by incubation with a secondary antibody coupled to a fluorophore, which emits a fluorescent signal detected by fluorescence microscopy. Using this approach, it has been shown that pericyte coverage was abnormal in mice that possess an invalidation of Platelet-derived growth factor (PDGF) in the retina.

Other in vivo models related to tumor angiogenesis involve the analysis of the vascularization during tumor growth in animals. Different models exist in this respect. There are tumor graft models where tumor cells are grafted in the animal and then tumor growth and angiogenesis are measured. In addition, there are transgenic mouse models that produce spontaneous tumors. Many of these models have been developed. I would only mention here one of most widely used models, namely the RIPTag mouse model developed by Douglas Hanahan (University of California, San Francisco, and now ISREC, Lausanne) in 1989 [250] (Fig. 11.5). We have mentioned this model before, but present a little more detail here. A gene called large T antigen, capable of inducing transformation and tumor progression, is expressed in β-cells of the mouse pancreas. These mice develop endogenous pancreatic tumors, which begin with hyperplasia and then evolve toward a highly vascularized and invasive tumor. This model has a very precise angiogenic "switch" and it is, thus, possible to distinguish between islets that have undergone angiogenesis and those that have not. The model has allowed numerous studies on the nature of the angiogenic "switch" or of stimulating and inhibiting angiogenesis factors, as well as some of the mechanisms of resistance to anti-angiogenic molecules. For example, it has been shown with this model that blocking the VEGF pathway by inhibitors may result in the induction of alternative pathways such as the fibroblast growth factor pathway or pro-invasive mechanisms involving the c-Met receptor located on the tumor cells [182, 228].

It is evident that these studies involve animal experiments. Animal experiments in the angiogenesis field are essential and involve mice, rats, chicken embryos, or zebrafishes, and only in a few cases large animals such as pigs (essentially for study of cardiovascular angiogenesis). The legislation on animal experimentation has changed considerably, in particular with recent European directives, which take account of animal welfare and suffering (European Directive 2010/63/EU, revising Council Directive 86/609/EEC 24 November 1986 and applicable in France from 1 January 2013). We are far from the time of William Harvey where vivisection was practiced and where the animal was seen as a creature devoid of soul and therefore incapable of suffering.

What About the Clinic?

12

The list of pathologies in which an excess of angiogenesis is observed is relatively long (see Table 12.1). However, the importance of angiogenesis in the pathophysiology of these diseases varies. It certainly plays a preponderant role in cancer and neovascular eye diseases. However, even in cancer, the importance of angiogenesis seems to vary from one cancer to another and from one stage of the disease to another.

The pharmaceutical industry has long been undecided about the development of molecules interfering with angiogenesis. This is because endogenous inhibitory molecules such as angiostatin and endostatin have not lived up to their promise in preclinical and clinical trials.

However, Genentech, under the impetus of Napoleone Ferrara, developed a humanized monoclonal antibody that had excellent inhibitory effects in preclinical models [251]. Clinical trials were then conducted, and in 2004, in the renowned *New England Journal of Medicine*, a very important article was published that definitely proved that inhibition of angiogenesis was a promising strategy for the treatment of cancer [139]. Hurwitz and colleagues showed clearly that patients with colon cancer with metastasis treated with a combination chemotherapy-VEGF inhibitor had significantly increased progression-free survival. This inhibitor is called bavacizumab (Avastin^{TM}) and has been used clinically for several years in different cancers.

Another strategy was developed at the same time. This involved small chemical molecules that bind to the tyrosine kinase domain found in the intracellular domain of angiogenic factor receptors such as the VEGF receptor. These molecules often have a more or less narrow specificity, that is, they can bind to several receptors with varying affinities. Molecules include sunitinib (Sutent^{TM}), sorafenib (Nexavar^{TM}), or PTK 787 or temsirolimus. Sutent^{TM} and sorafenib preferentially, but not exclusively, bind the VEGF receptor. Sutent^{TM}, for example, can also bind FGF and PDGF receptors and bind to other molecules within a cell. The affinities are nevertheless lower for the latter substrates.

A. Bikfalvi, *A Brief History of Blood and Lymphatic Vessels*,
https://doi.org/10.1007/978-3-319-74376-9_12

Table 12.1 Angiogenesis in human pathology

Pathology	Excess of angiogenesis	Defect in angiogenesis
Cancer	+	
Retinopathy	+	
Age related macular dystrophy (AMD)	+	
Alzheimer's disease	+	
Lateral amyotrophic sclerosis (SLA, Lou Gehring disease)		+
Atherosclerosis	+	
Hypertension	+	
Crohn's disease	+	
Lupus	+	
Nephropathy		+
Chronic wounds		+
Coronary artery disease		+
Diabetic ulcer		+
Multiple sclerosis	+	
Vascular malformations	+	
Obesity	+	
Psoriasis	+	
Allergic dermatitis	+	
Kaposi's sarcoma	+	
Hypertension of the pulmonary artery	+	
Asthma		
Mucoviscidosis	+	
Intestinal inflammatory disease	+	
Parodontal disease	+	
Liver cirrhosis	+	
Diabetic nephropathy	+	
Arthritis	+	
Ovarian kysts	+	
Endometriosis	+	
Uterine bleeding	+	
Osteomyelitis	+	

How does anti-angiogenesis therapy work at the tumor level? It is surprising to find that most anti-angiogenic molecules are effective when they are administered together with chemotherapy. How then can we explain this effect? Two opposing concepts exist. In the first, defended by Robert Kerbel, professor at the University of Toronto, anti-angiogenesis acts as an additive to chemotherapy [252]. Anti-angiogenesis increases intra-tumor hypoxia, leading to tumor regression. In contrast, Rakesh Jain, a professor at the Massachusetts General Hospital in Boston, developed the concept of normalization of tumor vessels, which we have already briefly discussed [253] (Fig. 12.1). We know that, in tumors, vessels have an abnormal

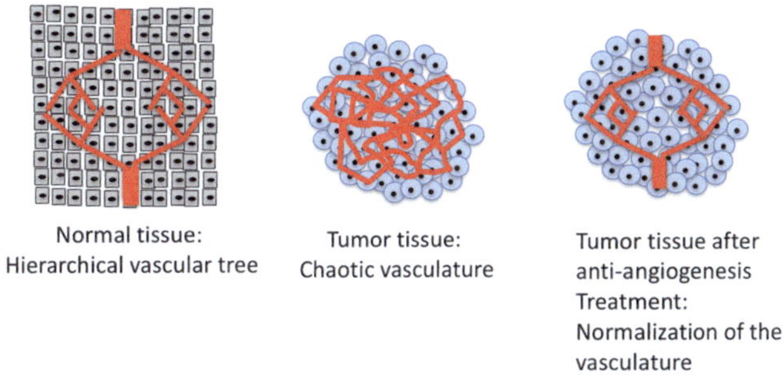

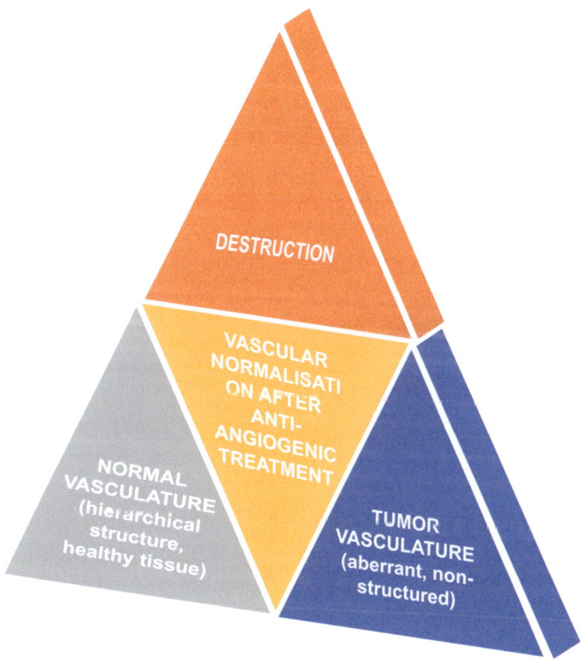

Fig. 12.1 Concept of vascular normalization. The tumor vasculature is abnormal in structure and functions. Anti-angiogenic therapy initially improves the structure and function of tumor vessels. However, aggressive or prolonged anti-angiogenic treatment may eventually destroy the vessels, resulting in a non-functional vascular system resistant to further processing. The dynamics of vascular normalization induced by VEGFR2 blockade (*top right*). Normal blood vessels in skeletal muscle (*left*); the following images show the vascularization of human colon carcinoma in mice (*middle*) at days 0 and 3 after the administration of specific VEGR2 antibodies (*right*). The figure is modified from Jain [255]. Schematic representation (*bottom*). Figures of the author. The photo was generously provided by Rakesh Jain

structure with a disturbed hierarchy and a faulty maturation. In this concept, anti-angiogenic therapy results in a return to a normal vessel structure as seen in healthy tissues. This concept is called "normalization." The vascular normalization concept is still widely debated and, as already mentioned, additional explanations exist. For example, some have thought that chemotherapy mobilizes hematopoietic cells from the bone marrow (which have a detrimental effect on the therapy) and this could be blocked by anti-angiogenic therapy [254]. Another explanation is the presence of VEGF receptors at the level of the tumor cells themselves and, thus, the anti-angiogenic treatment can also act directly on the tumor cells [228].

Another aspect is the escape of the tumor from the anti-angiogenic treatment. Greene already observed that some tumors develop, under certain circumstances, without a vascularization [64]. He wrote in 1938: "*Despite these conditions, the human tumor has grown in seven of the twelve rabbits used. Growth first became apparent toward the end of the third week. The extension of blood vessels from the iris into the transplant occurred in four of the animals between the thirty-fifth and fortieth days. Vascularization has not occurred to date in three of the animals in which primary growth was observed, but notwithstanding the fragments have continued to increase in size*."

Different mechanisms are responsible for tumor escape (Fig. 12.2). The first mechanism is the reactivation of angiogenesis by the induction of new factors and their receptors. The second mechanism is the activation of invasive properties directly at the level of the tumor cell. This involves the induction of molecules promoting invasion of tumor cells. The most important molecules are constituted by the cMET receptor and its ligand (factor which binds this receptor) HGF we have already encountered. Other additional mechanisms have been described and are discussed in this book.

Indeed, very interesting work has recently been done on this subject in the context of glioblastoma. Glioblastoma is a brain tumor with a poor prognosis. The current treatment consists of radiotherapy and chemotherapy, most often associating a molecule called temozolamide. The problem with glioblastoma is that it can modulate its dependence on vascularization. As indicated in a previous chapter, glioblastoma may have an "angiogenic" phenotype and an "infiltrative" phenotype. Given here are a few additional explanations as this point is important. In the "angiogenic" phenotype, many new vessels are present in the tumor. As for the "infiltrative" phenotype, there are no new vessels formed, but the tumor cells use the pre-existing vessels (normal vessels) to attach and to migrate as a train would use rails to advance. In patients, the angiogenic and infiltrative phenotypes are often mixed and this is variable. Glioblastoma can be experimentally manipulated to acquire an angiogenic or invasive phenotype and this can be used to reveal how these processes proceed on a molecular level. It has been shown that human glioblastomas implanted in animals initially have an invasive phenotype and may acquire an angiogenic phenotype after several passages [256]. Additional mechanisms, which involved the cellular stress response or CXCR3/LRP1 cross-talk as well as the role of the EGF receptor in the invasive process, have already been discussed in this book. Furthermore, Rolf Bjerkvig's team has described that metabolic adaptation occurs after anti-angiogenic treatment with Bevacizumab [257]. Indeed, an increase in glucose incorporation is observed in treated tumors with an increase in glycolytic activity and reduction in oxidative metabolism. Is this adaptation linked to the infiltrative phenotype and does this adaptation involve glycolysis when its migration is stimulated? This is in

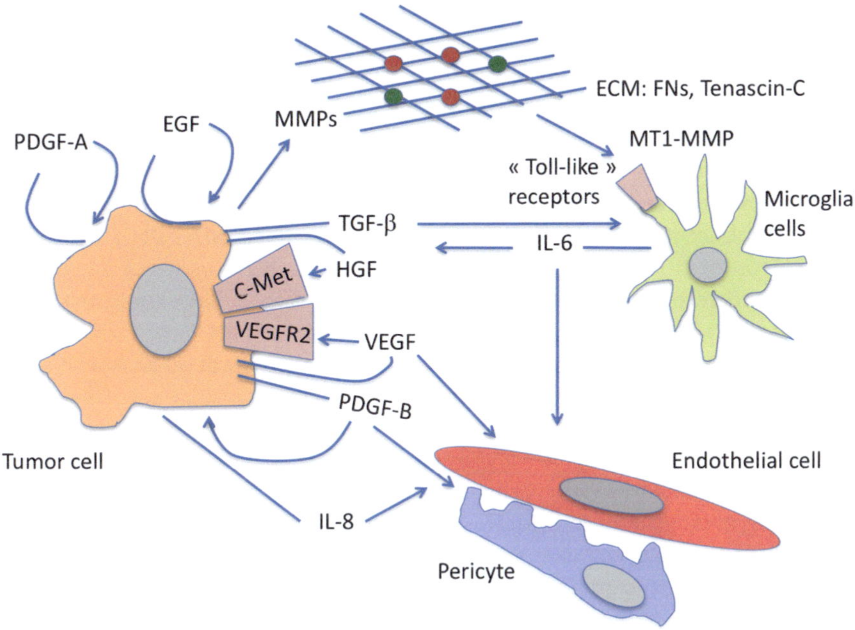

Fig. 12.2 Mechanisms of tumor invasion in brain tumors. Tumor cells interact with vascular cells, microglial cells, and components of the extracellular matrix. These interactions involve a variety of soluble regulatory factors and membrane receptors. On tumor cells, five growth factor systems are involved – PDGF/PDGFR, EGF/EGFR, VEGF/VEGFR2, TGFβ/TGFβR, and HGF/c-Met. At the level of vessels, various factors (VEGF, PDGF, IL-8, IL-6) are acting. In addition, microglial cells are activated by Toll-like receptors, membrane proteases (MT1-MMP), and matrix components (Fibronectin, tenascin). This leads to the migration of tumor cells into the cerebral parenchyma. The nature of pro-migration factors in the different structures of the cerebral parenchyma (vessel, nerve fibers, interstitial brain tissue) is not well-understood. Bradykinin is one of the only factors identified to date to attract tumor cells to blood vessels and allow migration using vessels as rails for invasion. Figure by the author reproduced with permission (Javerzat S, Godard V, Bikfalvi A. Future Neurology (2013) 8(2), 159–174. ISSN 1479-6708) [239]

apparent contradiction to the observation that oxidative metabolism is stimulated during invasion [258]. It is true that the latter observation was made in breast cancer, but one can wonder about these divergent results. Adaptation may be specific to the cellular or tissue context, but it would be better to have a more general explanation. In any case, no general conclusion can be drawn at the present time.

What is absolutely remarkable is that angiogenesis is completely blocked when tumor cells acquire an invasive phenotype. The vessels are tetanized by the assault of tumor cells.

Another target for clinical development was PLGF [172, 173] but this was abandoned because of conflicting results and failure in clinical studies. Indeed, based on the work of Peter Carmeliet, of which we have spoken previously, an anti-PLGF antibody was developed which had excellent effects concerning the inhibition of angiogenesis in some angiogenic tumors in an experimental setting. Nevertheless, conflicting results have been published which show either the absence of an effect or, in contrast, an acceleration of the tumor growth when the PLGF is inhibited and an increase of resistance to anti-VEGF treatment [174, 175].

Nevertheless, in eye disease, the results with regard to the inhibition of angiogenesis were spectacular [259]. Indeed, the use of anti-VEGF had very convincing results in the treatment of age-related macular degeneration (AMD). For ophthalmological use, Lucentis™ was granted marketing authorization (MA) [260]. Nevertheless, the cost of this treatment is very high and much higher than that of Avastin™. However, more recent studies have shown that Avastin™ has exactly the same efficacy as Lucentis™ [261]. It is therefore possible to treat these patients with Avastin™ rather than with Lucentis™ to reduce the cost. The disadvantage of these treatments is that these drugs must be administered by intraocular injection. Alternative strategies should be developed that use a more manageable mode of administration.

As mentioned above, various clinical trials have been conducted with the aim of stimulating angiogenesis in cardiovascular diseases and only mixed results were observed. In this case, cell therapy seems to be the more effective (see Chap. 9).

Another way of exerting a therapeutic effect on vascularization is vascular targeting. The idea of this approach is that it is more important to target a pathological vascularization than to target a function. Anti-angiogenic approaches aim to interrupt the function of an angiogenic factor or receptor, which is the case for all the anti-angiogenic treatments to date. Avastin™, for example, blocks the function of VEGF, as do VEGF receptor inhibitors (sunitinib, sorafenib, etc.). The "targeting" approach is quite different. The function is not important. What is important is that the marker, which can be a molecule that has no proven function, is expressed in a pathological vessel. This marker serves just as a recognition molecule, which is targeted by a suitable chemical (peptides, antibodies, etc.) coupled to cytotoxic drugs, vascular destruction agents, etc.) bearing a label that can be attached.

Philippe Thorpe carried out pioneering studies by coupling an antibody directed against activated endothelial cells to tissue factor, a molecule that induces blood coagulation [72]. Tissue factor is an activator of the extrinsic blood coagulation pathway leading to the thrombus formation made of fibrin and platelets. As we have seen, Paul Broca had proposed galvanopuncture, which aims to obliterate vessels and no longer allows the passage of fluids and nutrients. This is similar to what Philippe Thorpe did a century later using molecular tools that specifically attacked the tumor vasculature.

There are two distinct approaches to targeting (Fig. 12.3). One is using small peptides, the other using antibodies. Using the phage display strategy, Erkki Ruhoslahti and Renata Pasqualini (Burnham Institute, San Diego, USA) were able to identify labels specifically recognizing different types of endothelium [262, 263]. Thus, they were able to identify small peptide sequences that recognize only tumor vessels but not normal vessels. In addition, Erkki Ruhoslahti was able to identify peptide sequences that bind specifically to lymphatic vessels. The respective receptors of these peptides have been partially identified. It is evident that these

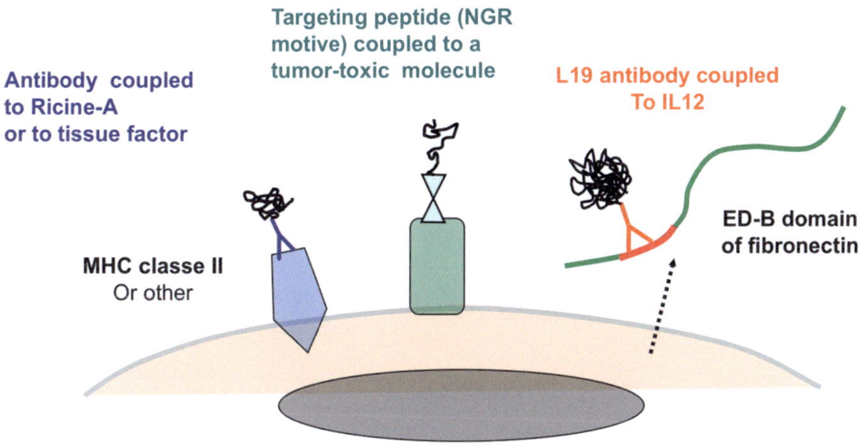

Fig. 12.3 Principles of vascular targeting. Figure by the author

peptides can be used for therapeutic and even diagnostic purposes by coupling them to an anti-tumor agent (for therapy) or to a tracer (for diagnosis).

Another approach is to use antibodies whose specificity for tumor or pathological (inflammatory) vascularization is proven. These antibodies can be obtained using the antibody phage display method. This was done by Dario Neri (ETH, Zurich) [264], who was able to identify antibodies specifically recognizing the EDB domain of fibronectin. This antibody has been coupled to various cytotoxic agents, TNF-α, or radioisotopes. The antibody is currently in clinical development and has shown promising results.

13 Lymphangiogenesis Enters the Dance

The lymphatic system differs significantly from blood vessels because it consists of a unidirectional system of tubes. The lymphatic vessels drain tissue fluids to the major lymph nodes and lymphatic vessels, which then reach the circulation.

As we have seen previously, the origin of the lymph vessels in the embryo was controversial. In particular, the centripetal theory opposed the centrifugal theory proposed by Florence Sabin [82, 83], who postulated a venous origin of the lymphatic vessels. This debate was finally decided in 1999 when Guillermo Oliver, at that time a researcher at St. Judes Children's Hospital in Memphis, Tennessee, identified the prox-1 gene (prospero) in the major embryonic vessels and clearly stated that this gene was involved in a cell engaging in lymphatic differentiation [265].

In the now generally accepted theory, lymphatic vessels first form at the level of the cardinal veins under the influence of specific lymphangiogenic factors. However, it seems that the cardinal vein is a heterogenous structure where only selected cells that are part of the inductive niche are involved [266].

As already mentioned in Chap. 2 "History of the Vascular System", recent results also indicate the existence of a non-venous contribution to lymphangiogenesis. Indeed, a significant contribution of non-venous-derived cells to the dermal lymphatic vasculature has recently been demonstrated, which is derived from a previously unknown lymphatic endothelial cell progenitor population [267]. Furthermore, it has been reported that cardiac lymphatic vessels in mice have a heterogeneous cellular origin and that the formation of cardiac lymphatic vessels is at least in part independent of sprouting from veins [268].

Kari Alitalo, professor at the University of Helsinki, had identified the most important molecule capable of stimulating lymphangiogenesis, VEGF-C [170] (Fig. 13.1). He quickly realized that VEGF-C had the ability to stimulate the growth of lymphatic vessels significantly. Kari Alitalo and Michael Detmar, a professor at ETH Zurich, both showed that VEGF-C plays an important role in tumor lymphangiogenesis [269, 270]. It was obvious that VEGF-C had to stimulate a

A. Bikfalvi, *A Brief History of Blood and Lymphatic Vessels*,
https://doi.org/10.1007/978-3-319-74376-9_13

Fig. 13.1 Kari Alitalo, the discoverer of lymphangiogenic factors. Image reproduced with permission of Kari Alitalo

specific receptor and this was indeed the case, as the VEGFR3 receptors were identified by Kari Alitalo's team [170].

What about other factors that can stimulate lymphangiogenesis [271]?

Among these, there is an intracellular protein called SOX18, which induces the expression of prox1. As we have seen above, prox1 is the master regulator of the differentiation of the lymphatic vessels. Prox1 induces another molecule, COUP-TFII, which is located in the nucleus of lymphatic cells and this results in lymphatic differentiation. An additional and significant number of receptors and signaling molecules also participate in lymphatic determination and lymphangiogenesis. For example, Neuropilin-2 (NRP2), ephrin B2, and Angiopoietin-1 (Ang1) are also essential in the formation of lymphatic vessels. Furthermore, the deletion of Angiopoietin-2 (Ang2) leads to a defect in the lymphatic vessels. Recent studies show that another protein known as CCG1 (collagen and calcium binding EGF domains 1 protein) is required for the sprouting of lymphatic vessels and the formation of the thoracic duct. In addition, it appears that activation of the intracellular MAP kinases (ERK) pathway controls the differentiation of veins toward lymphatic vessels during embryogenesis [272]. MAP kinases play a central role in the signaling pathways of most cells where they control cell proliferation and migration. It is surprising that this pathway is also crucial the differentiation of lymphatic vessels.

The interaction between the extracellular matrix and in particular between the collagen and lymphatic cells may enhance the dependence on the VEGFR3 receptor. Another reaction that also takes place during embryogenesis is the separation of the lymphatic vessels from the veins. This is achieved when blood platelets come into contact with lymphatic vessels where they form aggregates around these vessels by binding to podoplanin expressed on lymphatic cells. Two signaling proteins, called Syk and SLP-76, are also expressed in hematopoietic cells and participate in lymphangiogenesis. After their formation, lymphatic vessels are matured and reshaped into a branched network that invades the skin and organs. The circulation ("trafficking") of white blood cells (leucocytes) is also regulated by the lymphatic system and involves other receptors such as the lymphatic endothelium-1 receptor (LYVE-1), which stimulates T and B lymphocyte migration in the lymph node, and sphingosine-1-phosphate, which promotes transmigration from the lymph nodes. FGFs, which we have already discussed in the context of angiogenesis, seem to play a role not only in the maintenance of the vascular system but also in the lymphatic system. In collaboration with Michael Detmar (ETH, Zurich), we have shown that lymph node cells express the FGFR3 receptor and that its expression can be induced by the master regulator of lymphatic determinism, prox1 [180]. Others have reported that the FGFR1 receptor was the essential receptor for regulating lymphangiogenesis [273]. There is therefore a controversy in this regard. FGF and VEGF-C also seem to cooperate in the regulation of lymphangiogenesis [273].

The lymphatic system plays an important role in adults and is involved in various pathologies. Mutations in genes involved in lymphangiogenesis have been described, and are the basis of hereditary lymphoedema. For example, lymph-edema-distichiasis syndrome is caused by a mutation in the FOXC2 gene, which, as we have seen, interacts with the calcineurin-NFATc1 system and stabilizes the lymphatic vasculature, allowing an adequate response to biomechanical forces [274]. This is important for the formation of the collecting vessels and lymphatic valves during embryogenesis. The pathophysiology of this disease is characterized by an abnormal development of the lymphatic valves and by a defect of peri-endothelial smooth muscle cells, intercellular adhesion molecules (connexins), and basal membrane.

Another example is Milroy's disease, which is characterized by a lymphoedema that occurs early in life. A cause of this disease is a mutation in VEGFR3, which inactivates the kinase domain of this receptor or of VEGF-C itself [275]. In preclinical studies, this deficiency may be compensated by the administration of an excess of VEGF-C.

In tumor development, metastatic dissemination is a crucial step. One distinguishes a so-called hematogenous dissemination (via the blood vessels) and a lymphatic dissemination (via the lymphatic vessels). There are different types of cancers that exhibit lymphatic dissemination, such as breast cancer or melanoma. It is important to detect metastatic dissemination early, which is done, for example, by examining the sentinel lymph nodes during surgery.

One question that had to be answered was whether lymphatic dissemination requires an active process from the lymphatic vessels or whether it is only passive. This issue was solved by different approaches [276]. Michael Detmar and colleagues analyzed a cohort of patients and showed that the presence of lymphatic vessels within the tumor was an important prognostic factor for patient survival [277]. Furthermore, a decisive experiment was carried out using the RipTag mice [278]. These mice produce a tumor in the endocrine pancreas that does not show lymphatic dissemination. Nevertheless, the overexpression of VEGF-C, which, as we have seen, is one of the most important factors during lymphangiogenesis, leads to the formation of lymphatic vessels in the tumor and to lymphatic dissemination. Thus, it

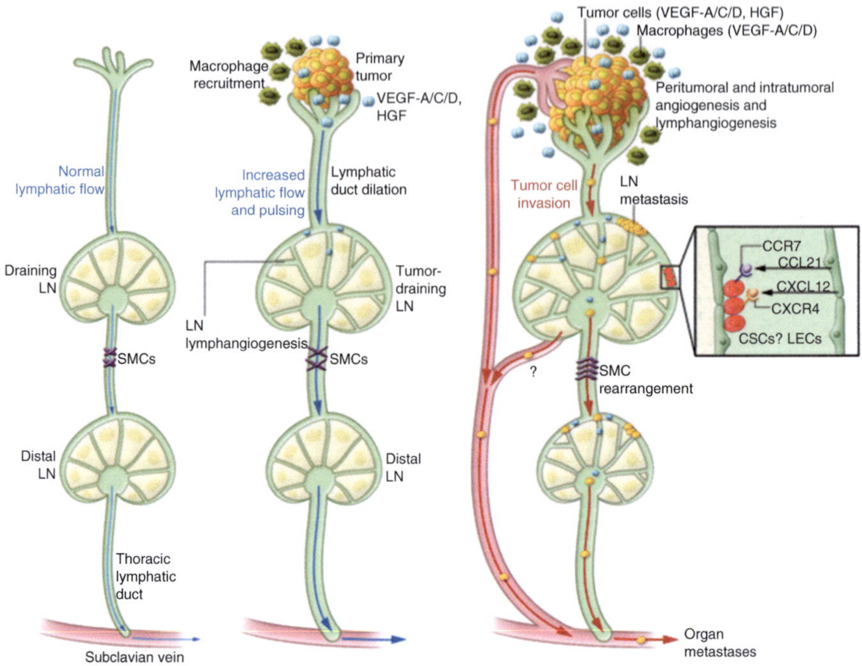

Fig. 13.2 Regulation of tumor lymphangiogenesis and metastatic dissemination. (*Left*) Normal lymphatic drainage takes place through lymphatic capillaries, lymphatic vessels, and collecting tubes, LN. (*Middle*) Lymphangiogenic factors produced by premetastatic tumors, including VEGF-C, VEGF-D, VEGF-A, and HGF, are transferred to peritumoral lymphatic capillaries and are transported by the collecting lymphatics to sentinel lymph nodes (SLN) where they act directly and induce lymphangiogenesis. Tumor lymphatic vessels are enlarged in size and have an increased conduction rate of lymphatic fluid. (*Right*) Once metastatic tumor cells have migrated to their draining lymph nodes, they are a major source of lymphangiogenic factors. This promotes the remodeling of the lymphatic vessels and lymph nodes and allows the formation of secondary metastases. SCC, cancer stem cells. Chemokines CCL21 and CXCL12, released by activated endothelial cells (ESL) in GS, could promote a niche of cancer cells with stem cell properties expressing CCR7 and CXCR4. Reprinted with permission: Karaman S and Detmar M, J Clin Invest. 2014, 124 (3):922–8 [276]

is now established that lymphatic vessels play an active role in metastatic dissemination. Figure 13.2 depicts schematically the regulation of tumor lymphangiogenesis and metastatic dissemination.

What About Stem Cells?

14

Takayuki Asahara and the late Jeffrey Isner, then director of a laboratory at the Dana Farber Institute in Boston (he died of myocardial infarction in 2001), had isolated circulating endothelial precursor cells using immunoabsorption with an antibody Anti-AC113 [279]. The AC113 antigen is expressed on a number of stem cells. When cultured, these cells acquire endothelial cell characteristics. Thus, peripheral blood contains progenitor cells capable of differentiating into endothelial cells.

Shanin Rafii and David Lyden of the Sloan Kettering Cancer Center in New York had postulated that stem cells produced in the bone marrow could accumulate at tumors and other sites [280]. This could have important consequences for tumor growth because 40% of the tumor vessels appeared to be endothelial cells from bone marrow progenitors. Their work also postulated that a cellular transcription factor, Id1, had an essential role in this process. Indeed, mice deficient in Id1 showed a defect in Endothelial progenitor cells (EPCs), which were in reduced number. On the other hand, angiogenesis was reduced and tumor progression as well as inhibition of micro- and macro-metastasis formation was observed. In the progression from micrometastases to macrometastases, Id1 seemed to act on another target gene called *p21*, which is inhibited by Id1.

Endothelial cell progenitor cells possess a number of markers to identify them. These include CD34, Flk-1 (VEGFR2), and CD-133. However, these EPCs appear to be a relatively heterogeneous group, and at least two different populations can be distinguished: an early, yet undifferentiated population that can produce paracrine factors, and late EPCs thought to have the ability to integrate into vessels under construction [281].

Two mechanisms have been proposed to explain the involvement of EPCs in angiogenesis. The first mechanism postulated the incorporation of EPCs into vessels during vessel formation. The second mechanism is characterized by the release of stimulatory factors from progenitor cells. Furthermore, it has been proposed that microparticles produced by activated cells or cells which have been subjected to stress or external aggression, as is the case in cardiac ischemia or peripheral tissue

A. Bikfalvi, *A Brief History of Blood and Lymphatic Vessels*,
https://doi.org/10.1007/978-3-319-74376-9_14

ischemia, could promote the differentiation of EPCs in order to promote neoangiogenesis [282].

Following the first description of EPCs, their identification and role in the vascularization of tumors has often been discussed [283]. A very critical article in this regard was published in Proc Nat Acad Sci USA in 2008 [284]. This article cast doubt on the significance of endothelial cell progenitors in tumor angiogenesis. According to these authors, EPCs derived from the bone marrow did not contribute to vascularization in tumors. For these authors, differentiation in mature and functional endothelial cells from precursor suspensions is a very rare event. This work has sparked great controversy and, in response, an "incendiary" article signed by, among others, Rafii and Lyden was published in Proc Natl Acad Sci USA. In their reply, they said: "*Purhonen et al. have refuted the data published in >50 reports, neglecting to quote key articles or utilize relevant models, and have drawn unsubstantiated conclusions about the contribution of endothelial progenitor cells (EPCs) to tumor angiogenesis that are not supported by their non-quantitative data and superficially executed experiments. Their study is flawed in experimental design and data interpretation*" [285].

Furthermore, clinical studies in patients with tumor recurrence [286] indicate that progenitors were present in vessels but at a low percentage (5% on average). These apparently contradictory data may be explained by the observation that the recruitment of progenitors into the tumor vasculature depends on the tumor grade. In addition, other non-EPC hematopoietic cells are important for the induction of a new vascularization and contribute significantly to it.

Another concept has also emerged more recently. This concept proposes that, in some tumors, tumor cells trans-differentiate into endothelial cells [287, 288]. It has been proposed that this trans-differentiation is carried out at the level of tumor stem cells. These cells are now called "tumor initiating cells" or TIC. Using the tumor model of glioblastoma, it has been proposed by some laboratories that TIC cells were capable of forming endothelial cells. These conclusions were based on the fact that endothelial cells of some of the vessels had markers that could be derived only from tumor cells.

As often happens in scientific research, disagreements arise and other laboratories dispute hypotheses and published data. Indeed, it has been reported that the population of endothelial cells with stem cell markers did not originate from the tumor cells [289]. In addition, another laboratory reported that pericytes were derived from tumor stem cells in a brain tumor model [290]. The authors reported, using lineage tracing, that a significant portion of the pericytes in the tumor were actually tumor cells. This laboratory, in turn, could not confirm trans-differentiation of tumor cells into endothelial cells.

There are some new publications that again give support to the trans-differentiation concept. Indeed, it has been reported that SCLC patients have rare Circulating tumor cell (CTC) subpopulations co-expressing markers of both endothelial and tumor cells such as vascular endothelial-cadherin and cytokeratins [291]. This is compatible with Vasculogenic mimicry (VM). Furthermore, higher levels of VM-like CTC are associated with worse overall survival, and tumor cells with both endothelial cell and tumor cell characteristics were also identified in experimental models.

A curious phenomenon called Endothelial-to-mesenchymal transition (EndoMT) has been recently described [292]. EndoMT involves the transdifferentiation of endothelial cells which lose their specific endothelial markers into unspecialized mesenchymal-like cells. These cells may then be further differentiated into other mesodermal cell types such as adipocytes, chondrocytes, or osteoblasts.

Again, the other way round, monocytes have been reported to undergo transdifferentiation into endothelial-like cells [293].

Taken together, the situation with regards to trans-differentiation is rather confusing where tumor cells trans-differentiate into endothelial cells, endothelial cells undergo EndoEMT, or monocytes undergo endothelial cell differentiation. Thus, in my opinion, for the time being, the results in favor of trans-differentiation, provided by some laboratories, are contradictory and remain to be confirmed.

15 The Neuronal Connection

In 1901, Christian Shiler [294] published an article in which he discussed the connection between the nervous and vascular systems. He wrote: "*In muscular and glandular tissues—and perhaps throughout the body: there is a vast peripheral nervous, plexus belonging to the capillary blood-vessels. These nerves of the capillaries, which may perhaps be regarded as nutritive nerves, regulate the production and transudation of the lymph, and are concerned in the mechanism of glandular secretion. They may influence, through their connections with the vaso-motor nerves on the arteries and veins, the blood supply to a part.*" He continues: "*Lastly, if the nerves surrounding a capillary are traced towards the center, they can be seen to pass into the plexus surrounding the larger vessels. Very frequently two fibers can be found running on the walls of the capillaries, and if one compares the nerve supply of striped muscle with that of the capillaries it can be seen that the latter are far more richly supplied with nerves than is striped muscle*"; and finally: "*What is the significance of these nerves of the capillaries, what is their function? Although I am not so fortunate as to have at my command a laboratory in which I could experiment on these nerves, yet if we take into consideration the histological facts together with certain clinical observations and physiological experiments which have been made on these nerves, although without any accurate anatomical knowledge about them, we have, I think, satisfactory evidence and scientific support for the hypothesis which I shall here state briefly: These nerves, so intimately connected with the capillaries, influence the protoplasm of their walls in such a way that, according to the activity of the nerves, the transudation of lymph is increased or diminished.*"

This article is important because it proposes for the first time not only an anatomical link but also a functional link between capillaries and nerves. With regard to the lymphatic capillaries, the nerves would have a role in controlling, in a certain way, the lymphatic vessels and, more precisely, the circulation of the lymphatic fluid. However, Shiler had described this interaction only in one direction: from the nerve to the capillary, not from capillaries to nerves. Moreover, he postulated that the nerves induce vasodilation of the lymphatic capillaries to allow

A. Bikfalvi, *A Brief History of Blood and Lymphatic Vessels*,
https://doi.org/10.1007/978-3-319-74376-9_15

the "trans-sudation" of the lymphatic fluid, which is inaccurate in light of our current knowledge.

Theoretically, the neuron-vessel connection can take place at several levels: (1) innervation of peripheral vessels, (2) vascularization of nerves, (3) role of vessels in neuronal activity, (4) role of vascular factors in neuronal activity, and (5) the role of neuronal factors in angiogenesis.

Functional interaction is also suggested because of many similarities between the nervous and the vascular systems. Both have what is called an afferent and efferent route, with the exception of the lymphatic vessels. The afferent pathway, in the case of neurons, consists of motor neurons, whereas the efferent pathway is constituted by sensory neurons. In the case of blood vessels, the afferent pathway consists of arteries and the efferent pathway of veins.

The two systems, vascular and nervous, are also interconnected and have common trajectories of migration during development. However, factors produced by neural cells and astrocytes clearly stimulate angiogenesis. The reverse is also true because vascular cells produce molecules such as artemin that attract axons to them [295].

If these two systems have such important similarities and interconnections, the factors or receptors described in the nervous system may have angiogenic functions and, conversely, angiogenic factors such as VEGF may have effects on neurons.

15.1 Neuronal Factors as Angiogenic Factors

Neuropilins play a role in the repulsion of axons and are located at the growth cone. Shay Soker and Michael Klagsbrun (Harvard, Boston) identified these neuropilins as a co-receptor for VEGF [296]. Indeed, binding to the VEGF receptor is stabilized by neuropilins. In addition, the team of Anne Eichmann (Collège de France, Paris) showed that NRP1 was expressed on arteries and NRP2 on veins [297].

In addition, the teams of Anne Eichmann, Dean Li (University of Utah, Salt Lake City), and Patrick Mehlen (University of Lyon) have also identified neuronal repulsive factors having a role in angiogenesis. One of the factors is the netrin-1. This netrin-1 has modulatory activities in angiogenesis [298, 299]. Another example is represented by a molecule called Robo4. Rob4 belongs to the family of semaphorin receptors involved in axonal guidance and repulsion. Indeed, it has been shown that Rob4 binds netrin on the endothelial cells and is involved in guidance [300].

Ephrins are another class of molecules involved in the nervous system and neuronal morphogenesis [301]. Ephrins are molecules that remain attached to the membrane and interact with receptors called Eph. The involvement of these ephrins takes place at two levels. First, they are involved in the separation/differentiation between the arterial and venous systems. Second, they have a role in vessel sprouting.

15.2 Angiogenic Factors as Neuronal Factors

It has been shown that VEGF has very important activities in the nervous system and this by directly interacting with the neurons [302]. Indeed, VEGF is able to stimulate neuronal survival and neuronal stem cells. Moreover, it is able to stimulate neuronal migration. For example, in the spinal cord, VEGF stimulates the migration of commissural neurons and is responsible for the correct establishment of connexions at this level. Importantly, neuronal activity can also be stimulated. The activity of granular neurons is stimulated by VEGF. In our laboratory, we have shown that VEGF could also intervene in the modulation of motor neuron activity in the spinal cord [303].

Another important aspect is that expression of VEGF can be modulated by behavior. Indeed, it has been shown that VEGF can be increased in mice by stimulating learning and memory activity [304]. Thus, this would indicate that VEGF, acting through KDR, mediates the effect of environmental cues on cognition. This may be because of direct (on neurons) and indirect effects (via the vasculature).

15.3 Innervation of Blood Vessels

Small and large vessels are innervated. There are neuronal connections that take place with the blood vessels. Vascular innervation obviously has a role in the regulation of vascular tone in larger vessels (arterioles, etc.). However, the innervation could also have a trophic role and produce survival signals for the vasculature.

What are the mechanisms of this innervation? As discussed previously, it has been shown that, during development, artemin produced by smooth muscle cells was able to attract sympathetic nerves to the vascular wall [295]. More recently, the team of Anne Eichmann has shown that the Netrin-1 produced by vessels was involved in the attraction of the axons to the vessels and their innervation [305]. The team demonstrated that netrin-1 is produced by arterial Smooth muscle cells (SMC) at the onset of innervation, and that arterial innervation requires the interaction of Netrin-1 with its receptor called DCC, located on the sympathetic nerve growth cones. The blockade of netrin or its receptor leads to a severe reduction in sympathetic innervation of arterial vessels and to a defect in vasoconstriction.

It has recently been claimed that large coronary vessels are paths for distal sympathetic axons to reach the subepicardial layer of the dorsal portion of the ventricles [306]. These vessels therefore participate in cardiac innervation.

15.4 Vascularization of Peripheral Nerves

The sensory and motor nerves must also be vascularized to ensure their function. This peripheral nervous vascularization is not well-understood at the present time. However, it is of paramount importance, particularly in pathology and especially in the context of peripheral neuropathy observed, for example, during diabetes. Some of the mechanisms have begun to be better known. It appears that molecules of the hedghog family are involved [307].

Other molecules involved are angiopoietins, also of therapeutic interest. We have already encountered angiopoietin in the previous chapters. It has been shown that a stabilized version of Angiopoietin-1, called COMP-Angiopoietin-1, has the ability to stimulate vascularization of the spinal cord and peripheral nerves [308, 309]. This could find application in the treatment of alcoholic or diabetic polyneuropathies.

What Future for Angiogenesis? 16

We have seen that angiogenesis research, at the molecular level, was initiated by studying the interaction between cancer cells and blood vessels. These studies have helped us to uncover some of the molecular mechanisms and actors involved in controlling the growth of blood vessels. Knowledge from these studies has entered many other fields, including developmental biology and cardiovascular and inflammatory pathology, which have continued to enrich our understanding of angiogenesis and physiopathology. The angiogenesis field has recently made progress at an exponential level. Many mechanisms have been elucidated and angiogenesis has entered the clinical era. One wonders what remains to be discovered! We discuss in this section a number of critical issues that seem important.

16.1 Molecular Characterization of Vascular Heterogeneity for the Identification of New Molecular Markers

To gain deeper insights into the molecular composition of vascular cells, new research strategies must be applied to angiogenesis research, such as single-cell sequencing technologies. Christer Betsholtz's team [310] has taken a step in this direction with their recently published results on the molecular composition of mural cells and have provided numerous new candidate marker genes. They used Pdgfrb-eGFP and NG2-DsRed reporter mice and performed single-cell sequencing in double-positive cells. This led to the identification of 250 new pericyte markers such as Vitronectin (Vtn) and Interferon-induced transmembrane protein 1 (Ifitm1), which were subsequently validated using fluorescent in situ hybridization and immunohistochemistry. This opens up an entirely new area of investigation and constitutes a rich source for new hypotheses for future studies.

A. Bikfalvi, *A Brief History of Blood and Lymphatic Vessels*,
https://doi.org/10.1007/978-3-319-74376-9_16

16.2 Inhibition of Angiogenesis

In oncology, angiogenesis inhibition has led to treatments used in the clinic. Nevertheless, disturbing observations have been made more recently that seem to limit the use of VEGF inhibitors in cancer. The initial observations were very encouraging because clear clinical responses were observed. This is the case in cancer of the colon or kidney, for example. This was reflected in improved progression-free survival [139]. Recent analysis, including very broad analysis known as "meta-analysis" (global analysis of several clinical studies) has also been carried out but data are conflicting [311, 312]. Two examples are cited here. In one analysis it was found that overall survival was not improved and patients treated with bevacizumab (Avastin™) had significant side effects with increased mortality (death from cerebral hemorrhage, etc.) [312]. Nevertheless, this was variable regarding the type of tumor. For breast cancer the risk was very low, which calls into question the methodology and the validity of this type of study. In another more recent meta-analysis, bevacizumab treatment resulted in a significant effect on survival and response in advanced colorectal, lung, ovarian, and kidney cancer but not in breast cancer [311]. In addition, an effect on cognitive functions was observed in patients with glioblastoma and treated with Avastin™ in a recent clinical study (RTOG 0825) conducted by the MD Anderson Cancer Research Center [313]. An effect on synaptic plasticity after treatment with Avastin™ has also been observed in experimental models [314], which is consistent with the clinical results observed.

Moreover, in experimental models, the acceleration of the tumor cell invasion in healthy tissue was observed [252, 315]. This was seen in different models. In particular, it appeared that too short exposure to the inhibitor would promote the appearance of invasion. In this case, we speak of evasive resistance, which means that the tumor not only no longer responds to the inhibitor but also develops an adaptive mechanism which consists in acceleration of tumor growth.

Does this mean that the inhibition of angiogenesis has no future in the treatment of cancers? This question should certainly be asked, but we should not throw away the baby with the bath water. First, it appears that there are no reliable biomarkers to stratify patients into responders and non-responders. If patients who respond well to anti-angiogenic therapy could be clearly identified, this would improve the use of anti-angiogenic molecules in these patients. In addition, it appears that kidney cancer is a subtype that better responds to anti-angiogenic drugs because, in this case, the therapy can be used alone, without the addition of chemotherapy.

Current anti-angiogenic drugs target VEGF and its receptor in the first place. However, the molecular regulation of angiogenesis, as we have seen throughout this book, is complex and multiple actors are at work. Inhibition of these other pathways could be effective when inhibitors of VEGF and its receptors have failed.

However, a question for which we do not yet have a complete answer comes immediately to mind. Is resistance a general property of the inhibition of angiogenesis or of a particular pathway, such as that of VEGF? It is clear that escape from VEGF inhibition can induce other angiogenesis factors such as FGF and that inhibition of FGF may, in turn, result in tumor regression. However, the activation of the direct invasion of the tumor cells has also been observed as we have already discussed, and it seems that a molecule called c-MET is mainly involved.

My personal opinion is that this phenomenon of resistance/escape ultimately encompasses two things. A form of resistance dependent on angiogenesis and a form of resistance independent of angiogenesis. If the former can be overcome by the development of new inhibitors that target other mechanisms or aim to destroy vessels, the latter needs additional strategies. These strategies are mainly aimed at blocking the direct invasion of healthy tissue by tumor cells. It should be kept in mind that, even in this case, a vascular element is present. Tumor cells use pre-existing vessels to invade tissues. These pre-existing vessels serve as highways for tumor cells and allow their penetration into the tissues. Pre-existing blood vessels are somehow hostages of tumor cells. This phenomenon is called co-option. The co-option mechanisms (i.e., what makes tumor cells such as melanoma coopt and migrate on vessels in the brain) are not known at the present time. It appears that a specific and close interaction between vessels and tumor cells is established and that factors exist produced by vessels that are important for attracting tumor cells and for allowing migration to the healthy tissue. We have reviewed some of the potential factors earlier in this book. It may also be that factors such as netrins that attract sympathetic neurons to the vasculature are also involved in the attraction of tumor cells.

In ocular diseases, inhibition of angiogenesis has given very promising results. In some cases, however, resistance is observed. This implies that other angiogenesis-dependent mechanisms are present that must now be targeted.

16.3 Stimulation of Angiogenesis in the Context of Therapeutic Angiogenesis

Therapeutic angiogenesis in the cardiovascular context was received with great enthusiasm in the late 1990s when the concept of revascularization was introduced (for references, see previous chapters of the book). Encouraging results of the preclinical studies have been published in uncontrolled therapeutic trials, particularly in the treatment of patients with lower limb ischemia. The enthusiasm generated by these early studies was hampered by disappointing results in major therapeutic trials. However, the situation seems better for lower limb ischemia, for which some encouraging results have been obtained. The limited response may be caused by various factors such as limited diffusion, the gene transfer system used, or patient selection. In addition, combinatorial approaches combining several factors, or even cell therapy, could be envisaged to improve therapeutic efficacy.

Most clinical studies had shown a modest but significant benefit when bone marrow-derived cells or cells isolated from peripheral blood were used in the treatment of ischemic arterial disease. These results have led to the development of complementary strategies such as the pretreatment of cells with stimulatory factors, which allow better recruitment to the ischemic site and better cell survival.

It should also be noted that not only should angiogenesis be stimulated, but also vessels should be stabilized and matured to allow adequate blood flow. New biomaterials and encapsulation techniques may help to optimize the efficiency of these treatments.

16.4 Lymphangiogenesis

The blockade of lymphangiogenesis could find application in the treatment of metastatic dissemination. The recent development of new non-invasive in vivo imaging techniques may lead to potential clinical applications for early detection of metastases based on tumor-induced changes in lymphatic vessels. The identification of specific markers reflecting changes in lymphatic vascularization is also an important aspect. For example, one study reported a correlation between the expression of ESAM, a molecule involved in lymph node endothelial cell junctions, and the appearance of ganglion metastasis in head and neck squamous cell carcinomas and colorectal cancers [316]. The development of tracers, in particular infrared tracers, to visualize lymphatic vascularization also makes it possible to quantify the response to the therapy.

However, there are pathologies which, in contrast, can benefit from a stimulation of the lymphatic network. This is the case of lymphoedema, which is observed after lymph node resection or radiotherapy in patients with breast cancer, tissue repair that is deregulated in diabetes, or in patients with congenital lymphoedema. Lymphangiogenesis could be induced in this case to allow the resorption of lymphoedema. Another aspect is congenital lymphoedema, which is also an important therapeutic application. This lymphoedema is caused by a mutation in the VEGFR3 receptor gene. This lymphangiogenesis could be induced by pneumatic compression or by administration of specific lymphangiogenic factors, which are under development. Active research is under way to develop therapeutic molecules capable of stimulating lymphangiogenesis (screenings of chemical or natural compound libraries).

The other application where stimulation of the lymphatic vessels may be important is inflammation [317] (Fig. 16.1). Surprisingly, VEGF-C stimulation of lymphatic vessels has anti-inflammatory effects in several skin and joint inflammation models. These effects are probably because of a better elimination of fluids and extravasated inflammatory cells, but also because of the modulation of immunity by lymphatic vessels. The lymphatic vessels are thus emerging as important targets for the development of new therapeutic strategies for treating inflammatory diseases. In this context, it is interesting to note that some of the anti-inflammatory drugs currently used are potent stimulators of lymphangiogenesis. Recent results on cardiac lymphangiogenesis point to a possible new treatment strategy for heart disease by interfering with heart lymphangiogenesis [318].

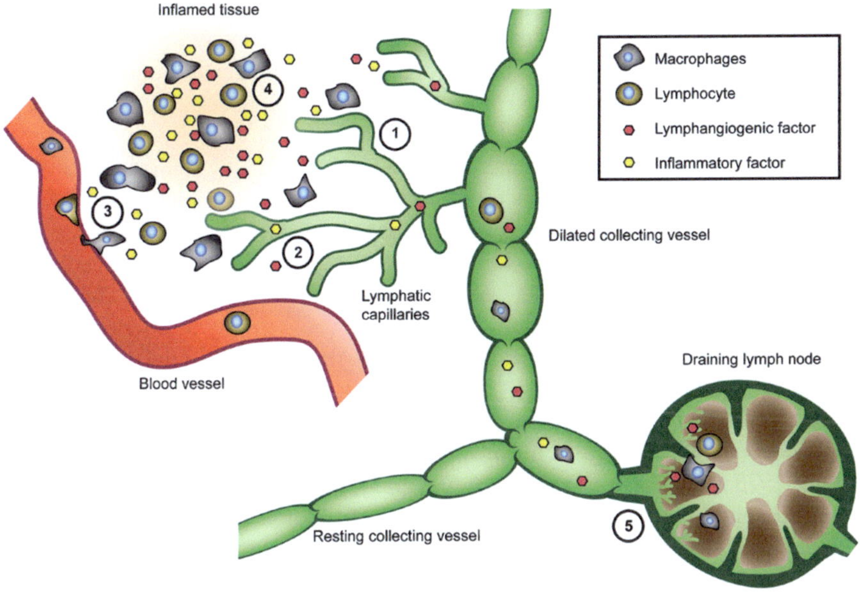

Fig. 16.1 Role of lymphangiogenesis in inflammation. Inflammatory cells accumulated in tissues are drained by lymphatics toward the lymph nodes. Stimulation of lymphangiogenesis could therefore lead to a reduction of the inflammatory reaction in tissues. Image with permission to reproduce, Springer Verlag (Dieterich LC, Seidel CD, Detmar M Angiogenesis 2014, 17 (2):359–371) [317]

16.5 Biomarkers and Angiogenesis

An important aspect that we discuss now is the problem of biomarkers related to angiogenesis.

A biomarker is a biological marker that reflects a physiological or pathological condition and can be used to monitor the evolution of a disease. Biomarkers are of various natures: biochemical factors (circulating angiogenic factors, metabolic factors, inflammation-related factors, etc.), circulating cells, histological markers, or markers derived from imaging.

Prognostic, predictive, pharmacodynamic, and surrogate biomarkers are generally distinguished. A prognostic biomarker provides information about the progression of the disease. The predictive biomarker is used to estimate the response to a given treatment. A pharmacodynamic biomarker changes its expression in the presence of a given treatment, which may reflect a response to therapy. A surrogate biomarker is intended to replace a clinical endpoint.

What is the situation in the context of angiogenesis? Several studies have been carried out, particularly in oncology (for more detailed information, readers can consult two excellent general reviews on this subject [319–321]). For instance, the degree of induced hypertension has been proposed as a predictive biomarker for the survival of patients treated with anti-angiogenic molecules. However, even if hypertension can be controlled by a suitable antihypertensive treatment, it does not seem appropriate to propose this clinical parameter as a biomarker because it is a treatment side-effect.

Regarding circulating angiogenic factors, VEGF was mainly studied. The results of the various studies are contradictory. Baseline values prior to initiation of therapy were correlated for some with progression (progression-free survival) in cancer, but in other clinical studies there was no correlation [322]. Moreover, the concentration of circulating VEGF tends to increase under anti-angiogenic treatment. The significance of this observation, however, is not clear. It could simply be a rebound effect under treatment with no particular significance for the evolution of the disease or the effect of the therapy.

Another parameter is polymorphism of the VEGF gene. Indeed, a Phase III clinical study with Avastin™ in breast cancer showed improved efficacy for the VEGF-2578 AA genotype [323]. However, this must be confirmed by other studies.

Other soluble factors, which have been measured, are represented by PLGF and soluble VEGF receptors [324]. These molecules appear to be increased after anti-angiogenic treatment. However, their value as a prognostic or predictive biomarker is uncertain. Inflammatory cytokines have also been measured and it appears that their increase in circulation is correlated with the appearance of more aggressive metastatic lesions. The value of circulating cells (endothelial progenitor cells of endothelial cells, inflammatory cells, etc.) as biomarkers is not clearly established.

A perhaps more promising aspect is the development of more efficient in vivo imaging techniques. Imaging techniques are based mainly on dynamic Magnetic resonance imaging (MRI) and Positron emission tomography (PET), and are intended to determine perfusion, blood flow, or vascular permeability. A recent development is vascular MRI (vessel caliber MRI), which measures Mean vessel diameter (mVD), mean vascular density (Q), and Vessel size index (VSI) in a tumor [325]. This method, which was introduced 20 years ago, has been greatly improved to result in the imaging of the vascular architecture called VAI (for vessel architectural imaging). This technology is not yet available in most clinical centers but deserves wide distribution. This could make it possible to extract parameters useful for monitoring the clinical evolution.

What is the situation regarding biomarkers for other angiogenesis-related diseases? Only cardiovascular diseases are discussed here. Cardiovascular diseases include a number of clinical entities: coronary artery disease ischemia and myocardial infarction, peripheral arterial disease (ischemia of the lower limbs, ischemia and cerebral infarction), ischemia of the retinal vessels (in the context of diabetes, for example), and heart failure. There are a number of conventional markers of cardiovascular disease such as LDL, HDL cholesterol, or creatinine kinase, but promising data have been published for troponin, natriuretic peptides, growth hormone, galectin-3, the soluble receptor of interleukin-33, ST2, and proneurotensin. If these markers are promising (troponin measurement is in clinical use), it is important to note that there are sex-related differences that may influence the diagnostic and predictive value of the assay [326]. Among these markers, growth hormone, galectin-3, and ST2 are related to angiogenesis and have stimulatory effects on vascular development.

What is the value of conventional angiogenesis factors such as VEGF as biomarkers in cardiovascular disease? As already indicated, genetic variants of VEGF have been detected, including some variants comprising Single nucleotide polymorphisms (SNPs). Only a few of these SNPs have been associated with cardiovascular disease. For example, SNP-634 C/G VEGF is associated with an increased risk of myocardial infarction and development of cardiac insufficiency [327]. The value as a predictive biomarker for the determination of VEGF in blood for cardiovascular disease is not clear.

An interesting factor for heart disease is VEGF-B [328]. VEGF-B has been shown to be highly expressed in cardiac tissue and has significant arterial effects. Its expression is decreased in heart disease. Decreased rates were measured and predict an increased risk of ventricular dysfunction after cardiac ischemia. VEGF-B could therefore be an interesting biomarker in the context of cardiac ischemia, especially as its predictive value seems to be greater than troponin T.

Another potential marker is chemokine CXCL4L1. It is a modulator of angiogenesis on which my laboratory has worked very hard. Reduced circulatory levels of CXCL4L1 have been shown to be associated with adverse outcomes in patients with preserved ventricular function after myocardial ischemia and as an independent prognostic biomarker [329].

A member of the family of fibroblast growth factors, FGF-23 is an independent risk factor for cardiovascular disease, particularly in patients with chronic renal disease [330]. The role of FGF-23 in angiogenesis is unclear and is considered to be a necessary factor in osteogenesis and in the regulation of vitamin D.

Other approaches include metabolomics studies in cardiovascular diseases, such as the measurement of Short chain dicarboxylacylcarnitines (SCDA), which shows a significant association with cardiovascular risk [331]. It is not established that SCDAs are derived from endothelial cells.

All these observations have been made in single studies and should be confirmed by wider studies.

16.6 MicroRNAs and Long Non-coding RNAs

Another recent evolution in the context of angiogenesis is microRNAs (miRs) and Long non-coding RNAs (LncRNAs).

MiRs constitute a class of non-coding RNAs 19–25 nucleotides in length and conserved. MiRs generally negatively regulate the expression of genes at the post-transcriptional level by translational repression or degradation of messenger RNA. In some cases activations have been observed. This has important implications for a variety of cellular processes. More than 2500 miRs are currently identified and more than half of the known genes are potential miR targets. What is the role of these miRs in angiogenesis? Studies have shown that a number of miRs have a function in angiogenesis. For example, miR 92a, miR-27b, and miR-130a stimulate angiogenesis whereas miR-221 and miR-222 are inhibitory.

Many factors can regulate miR in vascular cells. These include most of the factors we have studied in this book (VEGF, FGF, hypoxia, inflammatory cytokines, etc.). For VEGF, for example, miR 126, miR-296, and cluster mIR 17–92, which includes six miRs (17, 18a, 19a, 20a, 19b-1, 92a), are induced [332]. MiR 17–92 appears to be the most important in angiogenesis. The effects of miRs on angiogenesis may also be indirect, as is the case with miR 200 which targets interleukin-8 and chemokine CXCL1 and inhibits their expression [333].

These miRs are of therapeutic interest in diseases related to angiogenesis and future developments should offer interesting prospects. Therefore, inducing or inhibiting miRs may be of great interest to the therapy of cancers and cardiovascular diseases. These technologies, which include the use of antagomirs, LNAs (Locked nucleic acids), and PNAs (Peptide nucleic acids), decoys, mimetics, and precursor molecules, are being investigated to validate these strategies for future potential application in the treatment of these diseases.

MiRs also have potential interest in diagnosis, and various studies show this. For example, in clear cell kidney cancer, clinical studies of sunitinib resistance have been performed, which show a significant association of some miRs to the response to treatment by the determination of miR in blood [334]. For the cardiovascular field, many have been associated with coronary artery disease [335]. This includes 17 induced miRs and 8 miRs whose expression is decreased. Moreover, it seems that three miRs (miR-126, 197, and 223) have a predictive value in the occurrence of myocardial infarction in these patients.

These few examples are only an insight into the clinical potential of miRs for angiogenesis-related illnesses, and the reader may look further into the various references in this part of the book.

With respect to LncRNAs, I must be brief. The LncRNAs are non-coding RNAs of more than 200 nucleotides. These LncRNAs represent various entities, including antisense transcripts (complementary strand that does not encode a protein) of coding genes, or those localized between the coding regions of the genes. Their role is not yet clearly understood, but data indicate a function in regulating the expression of various genes by an epigenetic mechanism that involves chromatin remodeling, splicing regulation, and interaction with miRs to antagonize their effects.

Do these LncRNAs play a role in angiogenesis? A few observations demonstrate this. Indeed, the team of Stefanie Dimmeler (University of Frankfurt, Germany) showed that one of these LncRNs, MALAT1, had a function in angiogenesis [336]. It appears that MALAT1 has a biphasic effect on vascular development because its inhibition decreases proliferation but increases the migration of endothelial cells. The in vivo consequence is, nevertheless, an inhibition of angiogenesis when expression of MALAT1 is blocked. The significance of these data for physiology and pathology is still obscure and the mechanism of these effects is not yet understood.

More recently, MIAT (Myocardial Infarction-associated Transcript) has been shown to induce vascular dysfunction in diabetic retinopathy [337]. Inhibition of MIAT leads to vascular normalization and a reduction in the passage of fluids in the vessel toward the vascular leakage, and a decrease in the production of pro-inflammatory cytokines, which is accompanied by improvement of vision.

The study of LncRNAs is still in its infancy, but the potential for therapy and diagnosis of angiogenesis-related diseases is already apparent from these few examples.

Philosophy of the Vascular Tree

17

A number of aspects stem from the study of vascular development that are suited to a historical and conceptual analysis and are at present poorly investigated. Such an analysis could show how this scientific domain has developed and what meaning research conducted in this field has for science in general and for the development of scientific ideas or methodologies. We have already discussed some of these issues, and they are revisited in this chapter. However, I give herein a more conceptual perspective and my personal view on the subject. I focus on selected issues, including the analysis of paradigm changes, conceptual categories, cross-fertilization of fields, technological advances and impact on angiogenesis, evolutionary considerations, molecular proximity of different systems, and methodological considerations.

17.1 Changing Paradigms

According to Thomas Kuhn, a scientific paradigm is "*a universal recognized scientific achievement that, for a time, provides model for solutions of problems for a community of practitioners*" [338]. Kuhn makes the distinction between normal science that progresses by accumulation of data and knowledge, and scientific revolutions [338, 339]. Scientific revolutions occur when abnormalities in a research field are encountered that demand a full reconsideration of the conceptual framework in which science is conducted at a given time. There is a general tone of scientific relativism in the Kuhnian philosophy, but one can easily accept the Kuhnian scheme for scientific "progress" without falling into a relativistic posture. In Kuhn's view, a paradigm is the result of a radical transformation of a scientific field, which results in a new paradigm that conflicts with previous paradigms. In my opinion, one can adopt a less restrictive notion of *paradigm* and broaden its meaning. In this view, a paradigm is a sort of framework in which the working scientist carries out normal problem-solving science to fill up what is predicated by the paradigm. I would formulate the concept of micro- and macro-paradigms in this respect. Indeed,

A. Bikfalvi, *A Brief History of Blood and Lymphatic Vessels*,
https://doi.org/10.1007/978-3-319-74376-9_17

"revolutions" occur much more frequently on a smaller scale (micro-revolutions) and, in my opinion, they can be considered as micro-paradigms. These micro-revolutions do not affect the whole theoretical edifice of a scientific discipline but shed new light on and solidify the whole conceptual structure of a scientific theory.

Regarding vascularization, our knowledge has undergone a number of micro- and macro-revolutions leading to new micro- and macro-paradigms. A non-exhaustive list of these micro- and macro-revolutions is given in Table 17.1. It should be noted that these conceptual leaps have occurred over a period of more than 2,000 years!

The discovery of the circulatory system is seen as a macro-revolution that completely changed how we viewed the organization of living systems. The cell theory applied to the vasculature is another one. Yet another is the fact that vascularization is dependent on soluble factors produced by normal and pathological tissues, which led to the identification of these factors. Regarding micro-revolutions, I list some in the following: existence of attractive and repulsive factors, the postulate of the specificity of angiogenic factors, the discovery of Vascular endothelial growth factor (VEGF), the discovery of lymphangiogenesis factors, and the concept of guide cells ("tip").

I now briefly discuss in more detail the changing ideas about the significance of tumor angiogenesis. We already described the sequence of discoveries from a historical perspective in this book. Thus, historically speaking, the concept of the vascular Tumor microenvironment (TME) underwent three steps of conceptual modification: nourishing (Thiersch) → host defense (Goldmann) → tumor promotion (Folkman) → promotion and defense (present view) ("dialectic" progression) (Fig. 17.1).

It is important to note that Goldmann coined the notions of bodily reaction and host defense that are provided by the vasculature; this seems to be one of the first reports of inter-dependency between the vasculature and the immune system. As already mentioned, the tumor vasculature—immune interdependency, in the perspective of an anti-tumor response, was validated only recently by the identification of specialized vessels called high endothelial venules (HEV) in tumors, albeit prior reports indicate that immune cells report immune infiltration through the vasculature. These HEVs are present in some solid tumors, such as mammary carcinomas, and trigger an antitumor immune response by allowing the influx of Th1 cells, cytotoxic effector T cells, and naïve and central memory T cells into the tumor. It would be important to elucidate mechanistically how the number of these vessels can be stimulated to increase the therapeutic efficacy of immunotherapy, which has only be recently attempted by the Berger laboratory. This concept is a total conceptual inversion of the Folkman paradigm, because in this case tumor development is limited by increasing blood vessel growth (even if these are a specialized subtype of vessel) and thus therapeutic strategies must be developed to promote the growth of these vessels.

Another second conceptual inversion from the Folkman paradigm is the normalization concept as already mentioned in a previous chapter. As already discussed, in this concept, anti-angiogenic treatment is destined to kill aberrant vasculature in tumors and to normalize the morphology and functionality of the remaining tumor

Table 17.1 Non-exhaustive list of micro- and macro-paradigms related to vascular biology

	Reference and year	Micro- or macro-paradigms
1	Ibn Nafis (1242), Colombus (1559), Servetus (1553), Harvey (1628)	Existence of the pulmonary circulation. Nutrition and respiration are not independent in vertebrates but localized in the same circulation
2	**Harvey (1628)**	**The blood circulation in vertebrates is a closed circulation**
3	Harvey (1628)	It is not the diastole that attracts the blood, but the contraction of the heart (systole) that actively propagates the blood through the organism
4	**Schwann (1845, 1847), His (1865)**	**Vessels, in vertebrates, are composed of different layers and exhibits cells that are in direct contact with the blood (endothelial cells)**
5	**Hunter (1794), Goldmann (1907, 1908)**	**Vascularization is an active process in tissues**
6	Goldmann (1907, 1908)	The vasculature in tumors is part of a host defense mechanism
7	**Greenblatt and Shubik (1968), Ehrmann and Knoth (1968)**	**Soluble morphogenic factors are required for vascularization in tissues (tumor-derived factors)**
8	**Folkman (1971)**	**Vascularization is the prime ingredient of the integrated tumor ecosystem and essential for tumor growth**
9	**Gimbrone and Folkman (1972)**	**Blockade of angiogenesis halts tumor growth**
10	Ferrara (1989)	Angiogenic factors have vascular specificity
11	Keshet (1993)	Vascular morphogens act as gradients
12	Folkman (1971), Adams (2001), Eichmann (2004)	There are four types of factors: stimulatory, inhibitory, attractive, and repulsive
13	**Folkman and Hanahan (1996)**	**Tumors undergo an angiogenic switch to activate angiogenesis**
14	**Alitalo (1996)**	**The lymphatic circulation is independent of blood vascularization and has its own molecular mechanisms (lymphangiogenesis emerged as sub-field)**
15	Betsholtz (1999)	Mechanisms of pericyte recruitment to blood vessels
16	Gerhard and Betsholtz (2003)	Specific cells atop the nascent vessel guide the growth of the vascular tube (tip cells)
17	**Jain (2001, 2003)**	**Anti-vascular therapy normalizes the vasculature to allow better perfusion which improves the efficacy of chemotherapy**
18	Weinstein (2006), Affolter (2008), Lammert (2009)	Vessel lumen formation is an active process which requires at least one endothelial cell
19	Carmeliet (2004)	Angiogenic factors have extra-vascular properties

(continued)

Table 17.1 (continued)

	Reference and year	Micro- or macro-paradigms
20	Carmeliet (2013)	Vessel sprouting requires a specific metabolism
21	**Girard (2011)**	**Vessels may exhibit anti-tumor properties via the immune system**
22	Lammert and Melton (2001), Mastumoto and Zaret (2001), Keshet (2011), Rafii (2011)	Vessels have perfusion-independent roles by providing instructive signals to tissues

Macro-paradigms are indicated in bold

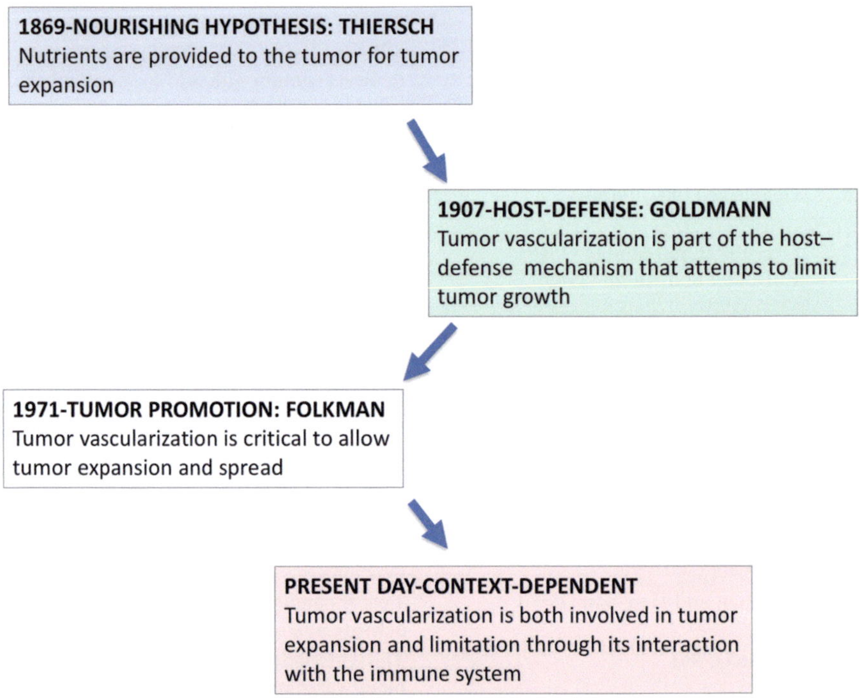

Fig. 17.1 Evolution of the concept of the vascular TME

vasculature. This has the consequence that blood flow in tumors is transitorily increased and this transitorily increases tumor growth but allows anti-tumor therapy to reach the tumor better and kill tumor cells.

A third conceptual modification is the fact that vessels not only seem to be channels for conducting nutrients and oxygen to normal and pathological tissues but also have a morphogenic function of their own by providing angiocrine signaling (see the Sect. 10.9 for more details).

Taken together, this places the vasculature at the center of multiple functions (Fig. 17.2).

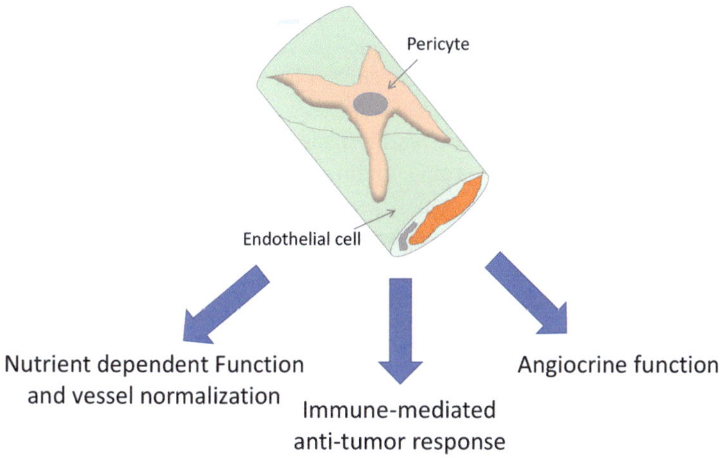

Fig. 17.2 The vessel is at the center of multiple functions. The function of the vasculature from the nutrient-dependent has been expanded to the normalization concept, immune-mediated antitumor response, and angiocrine function

17.2 Angiogenic Factors in Question

With regard to the concept of angiogenesis factors, there are two points to discuss – their discovery and their claimed specificity. Until the twentieth century it was elusive to think about diffusible factors that could control vascular morphogenesis. Vascular morphogenesis was instead viewed more to be the unfolding of an internal program within the vasculature itself. Several critical steps had to be taken in the discovery of morphogenic factors. The first was to demonstrate the existence of soluble and diffusible mediators of vascular development. The second was the characterization of the nature of these factors. The third was to gain mechanistical insights of how vessel development is regulated and the fourth was to attribute, if possible, a hierarchical and qualitative position to the various factors discovered. We have dealt with these in previous chapters of the book.

The discovery of soluble TAFs and growth factors raised the question about the quality or properties of vascular stimulating factors with regard to their multiple functions. These include growth promotion (vascular growth factors), migration (migratory factors), inhibition (negative regulators), guidance (guidance cues), and induction of permeability (permeability factors). These attributes have been determined "a priori," or can come from experimental evidence, or were imported from other scientific domains. "A priori" attributes are such as "if there are positive regulators there must be negative ones." Such a hypothesis may then lead to the search and discovery of these kinds of factors. Another way is experimental evidence where, for instance, a conditioned medium stimulates endothelial cell proliferation or induces vascular permeability, which leads to the identification of such stimulating factors. Yet another possibility is the import of a concept stemming from another field. The attribute "guidance cue" is such an example. Guidance cue is derived initially from developmental neurobiology and was later introduced into vascular biology (see Sect. 10.1 "Tip Cells").

It should be mentioned that discoveries made initially within different conceptual frameworks may lead to converging findings, as is the case for VEGF and VPF (Vascular permeability factor), which are identical molecules. The discovery of VEGF came from the idea to discover a factor that stimulated specifically the growth of endothelial cells. The discovery of VPF was derived from the idea to discover a factor that regulated vascular permeability.

It was believed at one time that angiogenic factors such as VEGF were specific for the vasculature and their only role was therefore to stimulate angiogenesis. The thinking was that a pleiotropic factor (i.e., that could interact with other cells outside the vasculature) was unlikely to play a predominant role in the morphogenic process. The concept that specificity equals importance was supported by in vitro studies and genetic experiments in mice where, for instance, VEGF knockout causes a very severe abnormal vascular phenotype. Thus, the wrong belief in the veracity of a concept can be reinforced by valid experiments and models. It is my view that such a kind of reductionist stand has been in the core of vascular biologists at the time and is still for many. Indeed, further studies discovered various other functions outside of the vasculature for vascular regulators. For example, VEGF has roles in the nervous system and the reproductive system. Thus, the quest for a "specific" factor, even if the assumption proved to be wrong, paradoxically significantly promoted research and allowed the discovery of VEGF and other angiogenic factors. This is an example of a conceptual error that can be, paradoxically, beneficial for research in a particular field of science. In my opinion, if one had made the initial correct assumption that no vascular specific factors exist, then already existing alternatives at hand would have satisfied scientist and hindered or slowed down the discovery of major vascular regulators. Fibroblast growth factors (FGFs), which are very potent angioregulatory molecules, were already discovered at that time. However, as the late Werner Risau pointed out at the time of the discovery of VEGF, FGFs were broad-range regulators and had not the "correct" spatial and temporal expression and hypoxia regulation, which, of course, VEGF possessed. To me, this was also the reason that pushed scientists to look for vascular specific factors.

17.3 Conceptual Categories Shaping Vascular Development Research

When focusing on vascular development, we can already define, in my view, several conceptual categories (Fig. 17.3) that include soluble angiogenic factors (module 1), vessel stability and maturation (module 2), sprouting and guidance (module 3), tumor/pathological angiogenesis (module 4), and effects of the vasculature on organ development (module 5).

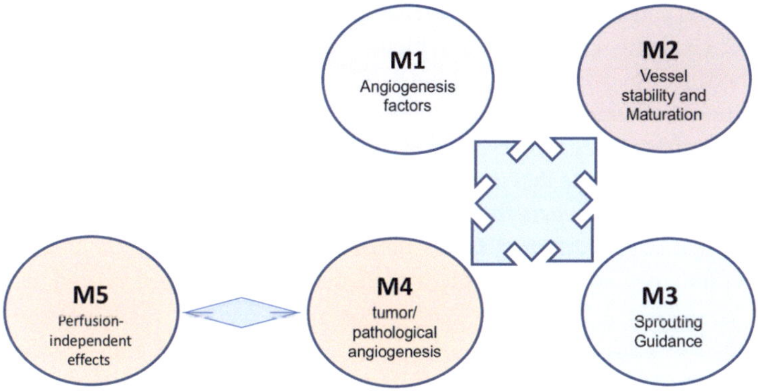

Fig. 17.3 Five conceptual categories (modules) in vascular development. There is interaction between four modules. Module 5 seems only indirectly conceptually connected

Module 1 is related to the discovery of soluble angiogenic factors. In this case, a first paradigm is represented by the contention that vascularization is an active process. The following sequence can be envisioned: vascularization is an active process → soluble factors are required for vascularization → vascular morphogens are specific → vascular morphogens act as gradients.

In module 2, the sequence is the following: vessels are composed of different cell layers → pericytes are present on capillary vessels → pericytes contribute to barrier function → pericytes stabilize capillary vessels and contribute to the response to angiogenesis factors or inhibitors.

For module 3 the sequence is: vessels are formed by sprouting → sprouting is dependent on morphogenic gradients → sprouting is dependent on tip cells → the orientation/direction of the vascular sprout is regulated by guidance factors.

For module 4 the sequence is: tumors are embedded into an integrated ecosystem → tumors are vascularized → vascularization is an active process in tumors with branches to go either toward a protumor (a) or antitumor (b) effect of the vasculature:

(a) Blockade of angiogenesis halts tumor growth → vascularization is dependent on diffusible factors → blockade of angiogenic factors halts tumor growth
(b) Vessels interact with immune cells → some tumor vessels have anti-tumor properties by promoting anti-tumor immunity → promoting these vessels inhibits tumor growth.

For Module 5, which is related to organ effects of the vasculature, perfusion dependent effects precede perfusion independent effects (perfusion dependent effects of the vasculature → perfusion independent effects of the vasculature).

It is important to remember that the progression of each of the paradigms is dependent on two factors of variable weight: the preexisting knowledge and the technological development at a given time.

Conceptual categories may also interact with each other (Fig. 17.2). For example, module 2 (pericyte coverage) and module 4 (tumor/pathological angiogenesis) interact because vascularization in tumors is immature and destabilized with a defect in pericyte coverage. There is interaction between modules 1, 3 and 4 because sprouting is aberrant in tumors and morphogenic gradients are abnormal. Module 5 is a stand-alone module that may only be indirectly connected to module 4 because emerging evidence connects this module to tumor angiogenesis. Thus, conceptual categories that depict different layers of knowledge have strong connections with each other and may contribute to the formulation of new hypotheses and paradigms.

17.4 Interactions Between Different Scientific Fields ("Cross-Fertilization of Fields")

17.4.1 Vasculature, the Central Ingredient of the Integrated Ecosystem in Tumors

It is surprising that cancer research opened the gate for a molecular explanation of vascularization. Oncology at the time of the pioneering work of Judah Folkman was under the influence of an exclusively tumor cell-driven explanation of cancer. Cancer research was mainly focused on changes taking place inside the tumor cell, such as alterations of the cell cycle or the study of oncogenes. The microenvironment was considered to be a non-active participant playing no role in the development of tumors. However, following the pioneering work of Judah Folkman and other researchers, it could not be ignored that the vascularization played a role in the regulation of tumor development. Folkman further coined the term "integrated ecosystem" where, in his view, the vasculature had the central role. This concept was instrumental in discovering many factors, receptors, and regulatory circuits that not only apply to the tumor context but also exhibit a general role in the vasculature. The concept of the ecosystem came from ecology and was introduced by Clapham in 1930 but it was Tansley who defined the concept fully and used it in 1935 in a publication [340]. Tansley devised the concept to draw attention to the importance of transfer of materials between organisms and their environment. Applied to cancer, this would mean that organs and tissues are seen as ecosystems in homoeostasis that tumor cells disrupt. However, this is not the meaning implied by Folkman when he speaks about tumors as integrated ecosystems. In his definition, tumors create their own ecosystem by re-educating the host and establishing privileged interactions with components of the micro-environment, among which, in Folkman's mind, the vasculature was playing the major role. This means that the tumor ecosystem disturbs the body's own homeostatic ecosystem by redirecting promoting signals to the tumor and by blocking inhibitory input from the body's ecosystem.

Why did Folkman call this an integrated ecosystem? Because he did not give any more details about this, one can only speculate about the reason. I give here my interpretation. The TME is composed of several entities that are metabolically active and have, in a certain sense, an autonomous component. In an ecosystem there is nutrient demand, waste production, and interaction of its constituents. Integrated may mean that all parts of it work together toward a common goal which, in the case of the TME, is to promote tumor expansion and spread. In Folkman's opinion, the vasculature is the essential ingredient of the integrated ecosystem because the vasculature is a crucial element in nutrition and waste exchange and this allows the passage of the tumor from a dormant stage to an actively growing and invading stage.

17.4.2 Angiogenic Factors Have Extravascular Properties and Vice-Versa

The history of VEGF's discovery is a striking example in this context. First, considered as a tumor angiogenesis factor, VEGF conquered developmental biology, cardiology, ophthalmology, neurology, and neuroscience, to cite only a few examples.

A number of molecules and receptors that were largely studied in other fields of the medical and biological sciences have been found to play crucial roles in the vasculature. As such, ephrins or netrins that play an important role in the development of the nervous system were discovered to exhibit important functions in the vasculature. These factors are not only molecules for axonal guidance but also instrumental for vessel guidance. Indeed, both vascular and nervous systems have an afferent and efferent system (arteries/veins, motor pathways/sensory pathways) and use similar molecular guidance systems. In the vascular system, there is the cell "tip" in the nervous system and the growth cone (see Sect. 10.1 "Tip Cells"). Furthermore, both systems seem interconnected at two levels. Vessels attract peripheral nerves and nerves attract vessels. In the central nervous system, a strong interconnection is found at the level of the blood-brain barrier. Hinman and Davidson proposed the term "Kernel" to designate a molecular regulatory circuit that is similar in different species but that comes in various flavors [341]. It may be not only the result of mutations or gene duplication – it may be composed of gene products, which have nothing to do with each other on a structural level. Moreover, this Kernel can function by transposition of a regulator assembly in another cell and tissue context. What is important in the Kernel is that its interrelations of the various molecular components are preserved.

In this context, we could eventually introduce the concept of micro- and macro-Kernel. A micro-Kernel could be the implementation of a specific function (such as the transposition FGF-Notch Delta in the tracheal system to the VEGF-Notch Delta in the endothelium) and the macro-Kernel could be a set of nodes which converge in a common functional purpose.

17.5 Technological Advances and Impact on Vascular Biology

One can identify different technological leaps that have accompanied the development of research in the vascular biology and angiogenesis field (Table 17.2). There are many techno-logical leaps such as, in the early days, injection of specific dyes to visualize the vasculature by John Hunter for the collateral circulation or the visualization of the capillary structure by Theodore Schwann. Another technology used in the early twentieth century was X-ray imaging that helps to visualize tumor vascularization in humans after bismuth oil administration, as done by Goldman in 1907.

One of the most important technological advances, in my opinion, is heparin-sepharose chromatography as many morphogens have a strong affinity for heparin. The interesting feature of this chromatographic method is that it allows the elution of a variety of factors using different ionic strengths. FGFs have a very high heparin-binding capacity whereas VEGF has a much lower affinity for heparin. Ferrara and collaborators used this method to purify VEGF, which started by the observation that a mitogenic activity was eluted from the heparin-sepharose column at a much lower ionic strength than FGF. This led ultimately to the purification and identification of VEGF. It should be noted that Dvorak and collaborators also used heparin-sepharose chromatography to purify VPF, which ultimately turned out to be identical to VEGF. Had this technology not been invented, it would have significantly delayed the discovery of vascular stimulating factors and slowed down the research of the entire field.

Table 17.2 Technological leaps in angiogenesis research

Technological leap	Reference	Consequences
Injection of dyes, etc.	Hunter (1794)	Visualization of in vivo angiogenesis
Microscopy	Schwann (1845, 1847), His (1865)	Visualization of vessel structure
Improvement of biochemical techniques (heparin-sepharose chromatography)	Singh and Klagsbrun (1984)	Purification of angiogenesis factors
Culture of vascular cells in vitro	Jaffe (1973), Gimbrone (1973), Buzney and Robison (1975), Campbell, Chameley-Campbell (1971, 1979)	Study of endothelial cell or vascular smooth muscle/pericyte phenotypes in vitro, elucidation of signaling mechanisms
Vascular specific animal models and gene deletions of vascular genes	Carmeliet (1996), Ferrara (1996)	Validation of an in vivo role of angiogenesis regulators, receptors, etc.
Molecular biology	Ferrara (1989), Ferrara and Williams (1992), Betsholtz (2016)	Cloning of angiogenic factors and receptors, micro-analysis of tissues at the gene level, single cell sequencing, etc.
Advanced imaging techniques	Jain (1987, 1992), Gerhardt (2010)	Visualization of fine morphogenic events (dynamically)

Additional important advances have been made by successfully isolating endothelial cells and other vascular cells from various sources and by successfully culturing them in vitro. This has allowed the investigation of cell phenotypes and behavior as well as cell signaling. Other no less important discoveries include vascular specific animal models that allowed specific deletions of vascular genes as well as molecular biology techniques that led to the cloning of vascular morphogens and receptors, as well as micro-analysis of tissues at the gene level. It is to emphasize that single cell sequence technologies are now applied to gain insights into the heterogeneity of the vasculature. Furthermore, recent microscopic techniques using two/multiphoton live imaging were instrumental in analyzing precise vascular morphogenic processes dynamically, such as tip and stalk cell organization in the nascent vascular tube.

17.6 Evolutionary Considerations and Principles

We have already dealt with this in a previous chapter of the book. However, I believe that it is useful to discuss some of the aspects here again.

The vasculature comes in different flavors in the animal kingdom. In invertebrates, open and closed vascular systems are encountered whereas in vertebrates only closed circulatory systems exist. In invertebrates such as *Drosophila*, the oxygenation system is constituted by the tracheal system, which is separated from the vascular system ("dual" mode). A comparison of the vascular system of *Drosophila* and vertebrates provides valuable insights into how the vasculature is organized. In *Drosophila*, there are coelomic cavities segmented by the dorsal parts of the mesoderm. The vascular system consists of a central contractile vessel ("heart") that receives hemolymph in the anterior part to be ejected at the posterior part. The wall of the vessel consists of mesothelial cells (also called myoepithelial cells because of their contractile ability) and matrix ("basal"). It should be noted that the matrix is located in the vessel lumen and is therefore exposed to the blood.

In vertebrates, the morpho-functional situation is quite different. First, the system is closed and allows the recirculation of blood through the venous and arterial systems. Second, an endothelium covers the internal surface of the vessel and is therefore in contact with the blood. In a pathological situation, the absence of the endothelium (because of damage caused by atherosclerosis, for example) is the initiator of adhesion and activation of platelets. In *Drosophila* this does not apply. In insects there are, most likely, circulating anticoagulant factors that prevent coagulation and maintain blood fluidity. Third, in vertebrates there is convergence of the functions of nutrition and respiration (oxygen). In *Drosophila*, the oxygenation system is constituted by the tracheal system, which is separated from the vascular system ("dual" mode). It should be noted that the hypoxia-inducible system in *Drosophila*, which involves HIFs, is located in the tracheal system and regulates FGFs. In vertebrates, the HIF system is located in blood vessels and surrounding tissue and is dependent on VEGFs. Altogether, these morpho-functional differences between invertebrates/insects and vertebrates show that, during evolution, a "shift" has occurred with regard to this function from the tracheal to the vascular system. As described previously, the term Kernel was proposed to designate control modules found in various biological systems, which exhibit different components, albeit having the same relationships with each other. Further insights can be gained from the study of *Botryllus schlosseri*, a marine invertebrate. As we have already seen in Chap. 3 "Evolution of Vascular System", in *Botryllus*, the endothelium is absent, but *Botryllus* has the ability to form external vascular tubular structures composed of epithelial cells. Surprisingly, these vascular structures express a homologue of VEGF receptor.

This raises the question of what the origin of the endothelium at an evolutionary level might be. Theoretically, both cell types, hemocytes and epithelial cells, could claim this role. Hemocytes are equipped with a homologue of VEGF receptor and are stimulated by VEGF counterparts. This suggests that hemocytes have acquired the ability to become/function as endothelial cells during evolution. This process must be accompanied by acquiring genetic elements of hypoxic regulation. We can infer from these observations that the function of the VEGF system was initially only aimed at controlling the movement of blood cells to the tissues. This function was hijacked during evolution to acquire a new morphogenic and structural role. It was no longer sufficient to convey the blood cells to tissues by an exclusive action on blood cells, but there was also the need to form the channels to achieve this effectively. When put into context, epithelial cells of *Botryllus schlosseri*, for instance, have gradually lost their dependence on VEGF, the role taken over by endothelial cells probably derived from hemocytes, which already have a dependency on VEGF for inducing cell migration. The alternative explanation is that the endothelium is derived from mesothelial or myoepithelial cells by trans-differentiation. This is not completely excluded because, as mentioned before, they may express receptors for vascular growth factors. However, this explanation seems unlikely because such a functional shift would imply a myoepithelial–endothelial transition. Furthermore, myoepithelial cells have a contractile capacity, which makes them cousins of pericytes and smooth muscle cells.

Do invertebrates have angiogenesis, strictly speaking? In vertebrates, angiogenesis is related primarily to the endothelial cells. They are the ones that initiate the cascade of events leading to the formation of the vascular tube. In invertebrates, these cells are absent. However, as we have seen, VEGF receptors are present in some invertebrates and this implies that VEGF is implicated in the formation of tubular structures. As already discussed, *Botryllus* have an internal and external circulatory system where epithelial morphogenesis is controlled by VEGF. However, the regulation of VEGF expression appears to be independent of hypoxia. Munoz and colleagues proposed the term "non-endothelial" angiogenesis for invertebrates [342]. However, more appropriate seems the formation of vessels in invertebrates as "vascular tubulogenesis" to differentiate specifically this phenomenon from "angiogenesis" which is associated with the endothelial organizing principle.

In summary, two important leaps occurred for the vascular tree during evolution: (1) a passage from the organizing principle "epithelium/myoepithelium" to "endothelium" and (2) a re-contextualization of regulatory mechanisms with the integration of the hypoxia control system into the endothelium and the surrounding tissue.

Related to this discussion is the concept of vascular mimicry, which may represent an ancient form of tube formation. It has been proposed that tumors, in some cases, produce vascular channels devoid of endothelial cells and only lined by tumor cells (see a more detailed discussion in the previous chapter).

17.7 Model Organisms and Angiogenesis

Angiogenesis uses a number of model organisms in research. These include the rodent model (mouse, hamster), the chicken embryo model, and the zebrafish model. In addition, research is carried out using the *Xenopus* (frog) embryo model. The model of corneal angiogenesis in rabbits was used to test the effect of activating and inhibitory molecules of angiogenesis.

Obviously, one may wonder how the results obtained in this way can be generalized. A simple test is to evaluate and compare the results obtained in different models and see whether they are consistent. In other words, this means that the results obtained on one model are the same in another model (or, better, in several other models). If, for example, we look at the VEGF family and its receptors, it has been clearly shown that they have very similar roles in the mouse, zebrafish, chicken embryo, and *Xenopus* models. In all these models the VEGF system is crucial to vascular development because extinction of the gene or its receptor(s) leads to a disturbance of angiogenesis with a defect in vessel sprouting. However, this rule must be applied with caution because in some cases it does not apply. During phylogeny, the molecular repertoire becomes more complex, and some molecules that play a role in mice and humans are not present in phylogenetically inferior organisms. An extreme example of this situation is represented by the variant-1 of Platelet Factor-4 (PF4v1), which is present only in humans, monkeys, and chimpanzees. It is therefore difficult to apprehend its true function because it would mean that its gene must be silenced in man or at least in the other two species. In this case, only the artificial expression of this gene in different organisms is possible and can indirectly inform us about its potential function.

Another situation is represented when two organisms have similar molecular systems but when the target tissue is different. Can we, in this case, deduce at least one common function? We have previously seen that the system regulated by hypoxia (together with the Notch and Delta systems) is involved in the formation of the tracheal tree in *Drosophila* and in the formation of the vascular system in vertebrates. However, in these two cases, these systems are involved in tracheal and vascular sprouting and therefore have similar morphogenic effects, even if some differences exist.

Another aspect is represented by the experimental approaches used in the different model organisms. Can they be transposed easily from one organization to another? For some approaches this is certainly the case. Developmental observation is common to any model organism. In addition, hypoxia conditions can be induced in mice and zebrafish, for example. Tumor angiogenesis has been studied in mice, chick embryo, and zebrafish. Other experimental approaches are rather specific. For example, if we want to study the role of the immune system in angiogenesis, it is preferable to use an immunocompetent and not an immunodeficient mouse model.

17.8 Scientific Methodology in Vascular Biology

The reductionist approach in the life sciences has produced spectacular results for the knowledge of living systems and has led to the development of treatments that have entered routine clinical practice.

Philosophers have extensively thought of how knowledge is acquired during the scientific enterprise and different theoretical frameworks have been formulated, which include inductive and deductive elements in varying proportions. These are inductive inference, deductive inference from hypothesis, the Bayesian approach, error-statistical approach, and by inference to the best explanation (for details on this topic see the excellent article by Marcel Weber (http://plato.stanford.edu/entries/biology-experiment/ with an extensive bibliography). These principles are, of course, general to experimental biology, but the reader may find it useful to discuss them in the present context.

In vascular biology, the principal method is inductive inference. Inductive inference obeys the following criteria: (1) simultaneous moment of occurrence of two events (if at any time an event takes place and at another time it does not and has all the conditions in common except one that exists only in the first event, the condition that is different between these two events is the cause or a necessary part of the cause of the observed phenomenon); (2) the two compared events must be uniform with agent/causal mechanism and the induced response need only be present in one but not the other situation; (3) other causes that can induce the response should not be present at the time of the experiment. Inductive inference is a key approach in vascular biology and angiogenesis research. It extrapolates from temporal occurrences a causal connection between the molecular alterations and the observed phenotype.

The following examples of inductive inference in vascular biology can be cited: discovery of VEGF and other vascular morphogens, stimulatory or inhibitory effects of factors that modify cellular phenotype or signaling, identification of vascular inhibitory factors, and loss-of function or gain-of function experiments for vascular morphogens, receptors or signaling molecules, etc.

In biology, there is not, strictly speaking, a similar theoretical framework as in physics, but there is a set of observations/assertions from which hypotheses can be deduced. As such, deductive inference is present in vascular biology, although this is not the "hypothetico-deductivism" which operates in other branches of science. An example of deductive inference is the angiocrine role of the vasculature. This is derived from the hypothesis that, besides the role in tissue oxygenation and nutrition, the vasculature and specifically the endothelium has a supporting role in normal and malignant tissue by producing growth factors, cytokines, or other cell-bound or soluble mediators. Another example is guidance (and guidance molecules), a concept derived from neurobiology with the formulation of the hypothesis that guidance also represents an important mechanism for directional growth of the nascent vascular tube.

It should be noted that inductive and deductive inference are closely linked and may flip back and forth. One can, for instance, start with a hypothesis and test the hypothesis by deductive inference, which may be followed by inductive inference, which can lead to new hypothesis that are again experimentally tested, and so on.

The Bayesian approach is a statistical approach that assigns a probability to an event related to another condition. Back in fashion, it found some interest in biology and in particular the analysis of genomic data, which is obviously important in angiogenesis research.

Another approach is the error-statistical approach. In this case, there is no probability assigned to hypotheses but the likelihood with which a hypothesis is tested is evaluated. In this case, the chance is evaluated that a negative hypothesis ("null hypothesis") passes the test. One can only suspect that this approach was inspired by Karl Popper and his falsification criterion [343]. In this case, a theory seems more solid if it has a significant risk of being falsified but after evaluation it is not. This has very general meaning and applies to all areas of vascular biology, but we can cite one example. The hypothesis that angiogenesis factors were specific for the vasculature had a high chance of being falsified, but it was not for a time because cell biology and genetics did not invalidate the hypothesis and, on the contrary, reinforced it. Thus, for a time it could be regarded as a very solid hypothesis. However, the hypothesis was ultimately refuted because more data were produced that demonstrated the contrary.

Inference to the best explanation (also called IBE) is an interesting approach for vascular biology. This approach bridges the gap between inductive inference, which is exclusively based on temporary occurrences to establish a causal link and mechanistic explanations, which involve concepts such as topology, structure, and feedback. How is IBE related to angiogenesis research? One example can be given here. By analyzing the structure of the VEGF promoter, the information derived from its structure identifies potential sites that are responsible for the binding of specific transcription factors, the role of which can then be functionally investigated through inductive or deductive inference.

We cannot follow the Popperian argument that the "queen" of scientific methods to acquire knowledge is the deductive method and the falsification procedure. Any researcher who has carried out experimental science knows that the process of knowledge generation in science involves a combination of induction, deduction, verification, and falsification. This is clearly illustrated by research in the angiogenesis field. Careful observation of the various morphogenic processes allowed the formulation of hypotheses, which were then validated by subsequent experiments. We must, however, be careful how much weight we give either to hypotheses or to experimental data. If a hypothesis is not verifiable or falsifiable, it can either be abrogated or included in what is called an auxiliary hypothesis that restricts the scope of the hypothesis and bring it into line with the experiments. This is a tricky situation because one can either over-evaluate the hypothesis or over-evaluate the experiment. However, it is only science when one can attribute the respective values (to hypothesis or experiment) and this can lead both to an enrichment of knowledge (if subsequent experiments by others are convincing and go in the same direction), and to drifts not conforming to the reality of the facts. A danger is always confirmation bias that can touch even entire research domains.

We have seen that, as with any research field, angiogenesis research has often taken tortuous paths, and is sometimes exposed to antagonistic concepts and encounters failures. We have encountered in the preceding chapters a number of these antagonisms and failures, to mention only angiostatin and endostatin in this respect. However, even failures can be beneficial, and all of this research contributes to the progress of science. There must be no church in science, no firm dogma, and every researcher, even the most reputable, must be able to be criticized and his theories tested, and this by the most novice scientists. This distinguishes it from the beliefs and ideologies for which full accession is required which cannot be questioned.

As in all human activities, science is not spared from certain drifts and can be contaminated by ideology and by using it in an indirect way in order to obtain a prestigious position or to increase its own power status. On the other hand, it sometimes happens that science undergoes what Irvine Langmuir called "pathological science" (https://www.cs.princeton.edu/~ken/Langmuir/langmuir.html). In a presentation to the Knolls Research Laboratory in December 1953, Langmuir describes the characteristics and symptoms of a pathological science which, to quote Langmuir in the text, are as follows:

> *The maximum effect is observed by a causative agent of low intensity and the magnitude of the effect is independent of the intensity of the causative agent. Intensity, and the magnitude of the effect are substantially independent of the intensity of the cause.*
>
> *The effect is of a magnitude that remains close to the limit of detectability because of the low statistical significance of the results.*
>
> *Claims are of great accuracy.*
>
> *Theories are always fanciful and contrary to experience.*
>
> *Criticisms are met by ad hoc excuses thought up on the spur of the moment.*
>
> *The ratio of supporters to critics is up to somewhere near 50% and then falls gradually to oblivion.*

In the scientific literature, a certain number of scientific works fall precisely into this category. Langmuir, who was a physicist, cites the Allison effect, the Davis-Barnes effect, N and mitogenic rays, extrasensory perception, and flying saucers. In the biological literature, the case of water memory was particularly revealing. Jacques Benveniste had published an article in the prestigious journal Nature which describes an experiment attempting to prove the principle of homeopathy [344]. This work had aroused much controversy and is now discredited, although Luc Montagnier, Nobel Prize in Medicine, who took up the torch of water memory, persists in this direction and affirms the opposite. There is even a recent publication on this subject in which the transmission of information by electromagnetic signals emitted by water, which had been previously in contact with the DNA, has been reported [345]! There is also a recent publication in a relatively modest toxicology journal by Seralini et al. [346] saying that tumors in rats can be induced by foods containing Genetically modified organisms (GMOs). This article had made a great impact in the media and the French newspaper "Le Nouvel Observateur" had published a long article with a title page concerning this work. I am not opposed to audacious ideas or a particular advocate of GMOs, but these works correspond to all the criteria Langmuir set out in his lecture and therefore reflects what he called "pathological science."

To quote the American astronomer Carl Sagan: "*Extraordinary claims require extraordinary evidence!*" Not everybody can call himself a convincing heretic with solid claims!

What about research on blood vessels? Can we detect signs of a pathological science? I have not found indications of a true pathological science in the various writings I have consulted. There are times when overly broad statements or erroneous conclusions based on biased experimentation have been made to cite only the angiogenin, angiostatin, and endostatin episodes that made a great deal of noise in the press of the time, or the role of endothelial cell progenitors in tumors.

However, everything that comes to mind does not detract from the fact that the basis of scientific activity is critical reason and not blind adherence to any postulate, and this makes science one of the noblest activities of the human mind.

Ill-intentioned minds may say that science does not need philosophy and can do very well by itself. However, conceptual research seems useful and can even enrich scientific research, as was shown by the philosopher Thomas Pradeu in his collaboration with the physician-biologist Éric Vivier on the immune system [347]. Such approaches deserve to be extended to other areas of biology, such as vascular biology or cancer.

17.9 Summary and Concluding Remarks on Conceptual Issues

We have identified several areas from the vascular development field that can be subjected to a conceptual analysis (Summary Box). These include paradigm changes, cross-fertilization of domains, technological development and the impact on discovery and knowledge building, evolutionary biology issues, and methods for scientific discovery in the vascular biology field. Paradigm shifts, as discussed, take place on a smaller (micro-paradigms) or larger scale (macro-paradigms). It is noteworthy that "false" paradigms can have a positive effect on the discovery path of a given scientific domain, such as the assumed specificity of vascular growth factors. One interesting aspect is how vascular development in tumors was viewed during history: first a nourishing tissue, then a host-defense mechanism, and finally a tumor promoter. Vascular biology has been cross-fertilized by other domains such as cancer biology or neurobiology. Cancer research has allowed the discovery of vascular trophic factors, and neurobiology has introduced concepts such as guidance and tip cells into vascular biology. Thus, concept transposition, such as seen for the notion of tip cells and guidance, has been/is a valuable device to fuel research in vascular biology and the angiogenesis field. Another important aspect is technology development and methods that are available at a given time, and how they impact on the formulation of concepts and theories. As shown, a simple biochemical method such as heparin-sepharose chromatography has led to the discovery of vascular development factors, confirming the anticipated "belief" of the existence of such trophic factors. This has then further spurred not only the vascular biology field but also many other fields, such as tumor biology and oncology. Finally, analysis of scientific methodology still shows a preponderance of inductive inference, even if deductive inference or methods such as IBE are also in use.

An important question is now where the field is heading and what might be the landscape in the next 20 years. I mention a few examples that, to me, seem important. At present, one important direction that is taken is the study of relation of the vascular system with immunology. The idea of Ernest Goldmann at the beginning of the last century has gained attention recently and, in this respect, recent work by two laboratories stresses the importance of HEVs to trigger anti-tumor immunity. It is important to stimulate HEVs in tumor cells in order to optimize immunotherapy. A step into this direction has been taken by demonstrating the combination of anti-angiogenic therapy and immunotherapy using anti-PDL1 can stimulate the number of these vessels in tumors. The effort in this direction must be enhanced.

Another area of investigation is how larger vessels are construed during development. Where do the endothelial cells that are incorporated into larger vessels come from? Is there retrograde trafficking of endothelial cells from capillaries and smaller vessels?

Important questions are also related to the molecular heterogeneity of vascular cells during development and pathology. Are expression profiles (such as obtained by RNAseq) of endothelial cells in one territory heterogeneous? What is the situation in pathology?

Computational biology has been used by a number of investigators to model vessel formation [348]. However, there is no convincing understanding unless a multilayer approach is developed with the integration of morphological and molecular data. This should be tackled in the future and the inclusion of new molecular data as indicated above can help the push in this direction. A step in this direction has been taken by the development of an open Library for Spatial Modeling of Vascularized Tissues (the microvessel Chaste) [349].

On the translational side, anti-angiogenesis therapy in cancer has deceived clinicians but, at the same time, was very successful in ophthalmology. Might anti-angiogenesis therapy in cancer still be used or abandoned in the near future? Industry is not investing additional efforts in the development of anti-angiogenic compounds in cancer. It is mandatory that new venues are explored, such as mentioned above with immunotherapy, for anti-angiogenesis therapy in cancer to survive.

Summary Box: Conceptual Issues

- Paradigm shifts occur at a micro- and macro-scale in vascular development research
- False paradigms may be useful in the discovery path and may lead to progress (i.e., VEGF specificity)
- Historically, concepts undergo transformations which may lead to opposing/conflicting views (i.e., the significance of tumor angiogenesis) with possible later resolution
- Cross-fertilization from other fields occurs. This happens at two levels, at a morphological level (i.e., the concept of tip cells and guidance from neurobiology) and at a molecular level (i.e., Netrin as a vascular guidance factor)
- Technology development of simple methods may have a major impact on the development of the entire field (i.e., heparin-sepharose chromatography)
- During evolution, the role of the various components of the vessel wall underwent modifications and the organizing principles were changed (i.e., myoepithelium → endothelium)
- Micro-and macro-Kernels can be defined. A micro-Kernel is related to the implementation of a specific function (i.e., transposition FGF-Notch Delta in the tracheal system to the VEGF-Notch Delta in the endothelium), whereas the macro-Kernel is related to set of nodes, which converge in a common functional purpose
- Analysis of scientific methodology still shows a preponderance of inductive inference, even if deductive inference or methods such as IBE are also in use. Inferences may flip back and forth (i.e., deductive ↔ inductive inference)

18 General Conclusion

Our journey in the company of blood and lymphatic vessel ends here. We encountered history, evolution, comparative biology, and the most advanced and current research in the field. We also discussed the philosophical aspects and the changes and evolution of concepts and paradigms. The specialist reader will hopefully forgive me for some shortcuts that cannot be avoided because this book is also written for a wider public. Nevertheless, I think I have covered the essential aspects concerning this subject. I have tried to use accessible language for a wider audience. Although sometimes a more technical language has been used, this has only been done to bring more precision to certain aspects of the ongoing research on vascular development.

I leave the last word to William Harvey, who contributed to the scientific revolution in the vascular biology field on which we are all dependent today. He wrote in the introduction to his famous book, *Exercitatio De Motus Cordis et Sanguinis in Animalibus*, the following:

> *My dear colleagues, I had no reason to inflate this treatise in a large volume by quoting the names and writings of the anatomists, or to demonstrate the extent of my memory, my knowledge, and my efforts; because I profess both to learn and to teach anatomy, not from books, but from dissections; not from philosophical theories, but from nature; and then because I do not think it is right or good to strive to take from the elders all due honor, nor to argue with the moderns, and to enter into a controversy with those who have excelled in anatomy and were my teachers. I would not like to accuse anyone who was sincerely concerned about the truth, or to accuse anyone who has fallen into error, of knowingly lying. I am a believer in the truth alone, and I can say that I have used all my efforts to try to produce something that should serve the good, be profitable to scientists and useful to the letters.*
>
> *Farewell, highly esteemed doctors, and kindly think of your anatomist,*
> *William Harvey*

A. Bikfalvi, *A Brief History of Blood and Lymphatic Vessels*,
https://doi.org/10.1007/978-3-319-74376-9_18

Erratum to: Evolution of the Vascular System

Andreas Bikfalvi

Erratum to:
Chapter 3 in: A. Bikfalvi, *A Brief History of Blood and Lymphatic Vessels*, https://doi.org/10.1007/978-3-319-74376-9_3

In the original version of Chap. 3, Fig. 3.2 was incorrect. The revised Fig. 3.2 is provided in this book version.

The updated online version of this chapter can be found at
https://doi.org/10.1007/978-3-319-74376-9_3

A. Bikfalvi, *A Brief History of Blood and Lymphatic Vessels*,
https://doi.org/10.1007/978-3-319-74376-9_19

Acknowledgements

I thank Anne Eichmann for a critical review of the text and the foreword she wrote, as well as all my colleagues and friends who read this book, in particular Curzio Ruegg and Markus Affolter for their comments, criticisms, and suggestions of the French version of the book. I also thank the members of my laboratory who have been with me for many years and have allowed me to conduct quality research in the angiogenesis and tumor biology field. My gratitude also goes to the various people who have had a considerable influence on my personal and professional development and who were, for a time, my mentors, in particular Adalbert Kemna, Jean Brière, Gérard Tobelem, Jacques Caen, Yves Courtois, and Daniel Rifkin. I also thank INSERM, Bordeaux University, the National Cancer Institute (INCA), the National Agency for Research (ANR), the Cancer Research Association (ARC), the League Against Cancer, and the European Commission for their support for the various research projects that have been carried out in the laboratory. Finally, I have special gratitude for my family, my parents Andras and Ildiko Bikfalvi, my sisters Barbara and Ildiko, and my children Alexis and Marianne, who were all on my side throughout the various trials of life.

A. Bikfalvi, *A Brief History of Blood and Lymphatic Vessels*,
https://doi.org/10.1007/978-3-319-74376-9

Appendix: Explanatory Note of Concepts and Technical Terms Used in the Book

3 R rule	Reduce, Replace, Refine. This is a rule established by Russell and Burch to improve the ethics of animal experimentation.
Actin	Major cytoskeletal protein involved in cell structure and mobility.
Adherent junctions	Junctions that also have junctions that are localized close to the apical pole. They are localized below the tight junctions and are, in case of endothelial cells, composed of Ve-cadherins.
Aneurysm	Malformation made of vascular cells. It is characterized by enlargement and formation of vascular sacs.
Angiogenesis	Formation of new blood vessels.
Angiogenin	Initially discovered in tumor extracts, it was thought to be the essential tumor angiogenesis factor. However, this molecule is actually a ribonuclease that degrades nucleic acids.
Angiopoietins	Molecules that play a role in the interaction between pericytes and endothelial cells. Angiopoietin-1 (Ang1) stabilizes this interaction, whereas Angiopoietin-2 (Ang2) has the opposite effect.
Antagomirs	(anti-Mirs or blockmirs) Synthetic oligonucleotides that prevent the binding of endogenous miRs to their target RNA sequences.
Anti-thrombin	An agent (molecule) with a capacity to inhibit thrombin (and its formation). Physiological anti-thrombins are antithrombin III and heparin cofactor II.
Antibody phage display	See "phage display".
Arteries	Vessels that transport oxygenated blood from the heart to organs and tissues.

A. Bikfalvi, *A Brief History of Blood and Lymphatic Vessels*,
https://doi.org/10.1007/978-3-319-74376-9

Astrocytes	These are glial cells providing support for different biological processes in the central nervous system. These cells can become transformed into astrocytomas (primary brain tumors).
Avastin	Made by Roche; commercial name of bevacizumab, which is a humanized antibody against VEGF.
Basement membrane	Membrane tissue comprising matrix components below the epithelial and endothelial cells in the case of the vessel. The major component of basal membrane in the vasculature is type IV collagen.
Blastema	A liquid or semi-liquid "living" substance capable of being transformed into solid elements.
BMPs "Bone morphogenic proteins"	Morphogenic proteins of bones. These are regulators of the TGF-β family that bind to the same family of receptors (ALK). There are several members of this family (BMP 1–15). They play an important role in angiogenesis and BMP9 is involved in Hereditary hemorrhagic telangectasia (HHT), also called Rendu–Osler disease, a hemorrhagic disease where endoglin is mutated and the Alk1 function is lost (co-receptor and receptor of BMP9).
Botryllus schlosseri	Marine invertebrates with an extracorporeal circulatory system.
Branchless/FGF	*Drosophila* fibroblast growth factor form.
c-Met	Tyrosine kinase receptor that is activated by Hepatocyte growth factor (HGF). This receptor is strongly involved in all phenomena related to migration and invasion of tumor cells. The activity of c-Met is regulated at the receptor level but not at the ligand level (HGF).
Carbohydrates (sugars)	Consisting of poly-carbon units linked together (Expl: triose, pentoses, hexoses ...) which can be cyclized.
CCBE1	Collagen and calcium binding EGF domains 1 protein. Protein binding collagen and calcium. It plays a role in the remodeling of the matrix and in cell migration. The expression is elevated in the ovaries. The mutation of its gene is associated with congenital lymphoedematous syndrome.
CD	Cluster of Differentiation. CD nomenclature has been proposed to classify antigens representing cellular molecules and receptors recognized by many monoclonal antibodies produced by different laboratories around the world. They usually carry numbers such as CD31, CD34, etc.

CD11b+Gr1+	CD11b is a white blood cell marker and labels monocytes, neutrophils and eosinophils. GR1+ includes different types of LY6G and LyGC markers. Ly6G labels the neutrophilic and eosinophilic mononuclear cells and LY6C the monocytic lineage. CD11b+GR1+ therefore includes monocytes, neutrophils and eosinophils.
Cephalochordates	The cephalochordates are characterized by the anterior prolongation of the notochord, without development of a brain. They have poorly developed sensory abilities. They live in the sea and feed themselves on organic particles.
Chemokines	Small regulatory molecules involved in various biological phenomena including inflammation, tissue repair, angiogenesis and tumor progression. They have a specific structure and are composed of monomers having a molecular weight of about 7,000 Da. These monomers are associated in multimeric complexes. They are composed of four major families, the chemokines CXC, C, CC, and CX3C. CXC, CC, and CX3C have two separate (disulfide) bridges with one amino acid in between in the case of CXC and three amino acids for CX3C. Chemokines C have a single disulfide bridge. Chemokines interact with different types of receptors (CXCR1, 2, 3, and 4 for CXC chemokines).
Chorio-allantoic membrane	Embryo membrane in the chicken egg through which gas exchange and waste disposal take place. It is a highly vascularized membrane and is used to test the angiogenic activity of stimulating factors or to explore the effect of inhibitors. It is also an ideal support for implanting tumor cells and fresh tumor pieces and monitoring their growth.
Coagulogen-Coagulin	Fibrinogen/fibrin-like proteins of invertebrates forming the clot through the conversion of coagulogen to coagulin.
Cocoonase	Enzyme of the silkworm that digests blood clots.
Coelum	Located between the outer wall and the intestine in invertebrates and is filled with liquid and lined with mesothelial cells.
Collagen	Protein of the extracellular matrix having a particular structure. Collagen is an insoluble fibrous protein in the extracellular matrix and connective tissue that is very abundant in the animal kingdom. There are 23 types of collagens assembled from 38 genetically

	distinct chains. Type 1 collagen is rich in glycines and prolines and consists of three helical chains (forming a helix). This tropo-collagen is then assembled to form collagen fibers.
Collateral circulation	These are blood vessels that bypass the obstruction in a larger vessel as is the case for example of a lower limb artery. These vessels are derived, in most cases, from the enlargement of arterioles by multiplication of vascular cells (arteriogenesis).
Contact spanning	Process of vascular extension in the retina which uses the tracks layed down by the astrocytes to develop. The vascular tubes therefore follows the pattern formed by the astrocytes.
Cooption	The ability of tumor cells to surround blood vessels and to use them as a support to migrate and grow.
Corpuscular theory	The nineteenth century theory that corpuscles (solid entities, cells, etc.) have the capacity to transmigrate through the vascular wall.
Cytoskeleton	The cytoskeleton is composed of structural and support proteins that are found in the cytoplasm between the nucleus and the plasma membrane.
Decoy	A nucleotide sequence more often of RNA type that acts as a competitor for the binding of various proteins (transcription factors, etc.) to the regulatory regions of a target gene. Their interest is potentially therapeutic.
Diploblast	Organism with two embryonic layers.
DMEM	Minimal essential medium modified according to Renato Dulbecco. It is a culture medium rich in various elements and nutrients in which many cell types can be grown. Calf or human serum is added to enrich the medium to allow cell survival and propagation.
Drosophila	Fruit fly.
ECGS	Endothelial Cell Growth Supplement.
EDB	Isoform of fibronectin generated by alternative splicing that is overexpressed around tumor vessels and in vessels in chronic inflammation.
Endoplasmic reticulum	An intracellular organelle that is located at the beginning of the protein secretion pathway. This is the entry route to the secretion pathway of polypeptide chains.
Endothelial cell map	Mapping of all intracellular molecular interactions occurring in an endothelial cell.

Endothelial cells	The main vascular cell that is in contact with the blood. This cell is characterized by the presence of specific secretory granules called Weible–Palade bodies and surface molecules (CD31, CD34, etc.).
Endothelium	Name that comprises the endothelial cell lining in the body.
EPC	Endothelial progenitor cells. Precursor cells giving rise to mature endothelial cells.
Ephrins	Ephrins are guidance molecules in the nervous and vascular systems. They are involved in guiding nerve fibers and vascular tubes. Ephrin ligands interact with Eph receptors. However, both ligand and receptor are anchored into the cell membrane. The interaction can be homotypic (same cell type) or heterotypic (localized on two different cells, one expresses ephrin the other the receptor).
Extracellular matrix	Extracellular fibrous deposit composed of molecules of very different nature (collagens, fibronectin, laminin, etc.).
Exudate	Liquid effusion in a tissue or cavity of the body.
Fenestrated capillaries	These are capillary endothelial cells with small pores. This makes them very permeable. They are found in various tissues where a high permeability is required (renal glomeruli, choroid plexus in the brain, small intestine and endocrine glands, etc.).
Fibroblast growth factors	Polypeptide growth factors with important regulatory roles in development and physiology. There are 23 members described for this family. These factors interact with four receptors (FGFR1–FGFR4) in a variable fashion.
Fibroblasts	An essential cell in the stroma producing matrix proteins.
Fibronectin	Multifunctional extracellular matrix glycoprotein with a dimeric structure. This molecule interacts with integrins (more precisely with integrin $\alpha5\beta1$ which are receptors integrated into the membrane; more specifically with integrin $\alpha5\beta1$ which are receptors integrated in the cell membrane). FN binds integrins by means of the RGD motive (arginine, glycine, aspartic acid).
Functional genomics	Study of the expression, regulation and function of genes.
Galvano-puncture	An ancient method introduced in the nineteenth century where electrodes are implanted in tumors and an electric current is applied.

Glioblastoma	Grade IV brain tumor, one of the most aggressive brain tumors. This tumor originates primarily from astrocytes.
Glioma	Primary brain tumor that may be of low-grade and high-grade. These tumors include astrocytomas and oligodendrogliomas. High-grade tumors are called glioblastomas.
Golgi apparatus	Intracellular organelles discovered by Camillo Golgi essential for the secretion of proteins.
Growth factors	A protein that can associate with a receptor and induce cellular signaling and cell growth and migration.
Hematopoietic cells	Cells derived from the bone marrow. These cells give birth to the different blood cells.
Hemocoel	Liquid contained in the coelum.
Hemolymph	Liquid contained in the internal space in invertebrates derived from the coelum.
Heparan sulfate (HS)	Molecule that is composed of sulfated sugar units that are present in the matrix and at the cell surface membranes. It is associated to a protein moiety to constitute a Heparan sulfate proteoglycan (HSPG). It modulates the activity of angiogenic factors.
Heparin	Glycosaminoglycan composed of sugars units which has a strong anti-coagulant activity. Heparin can also modulate the activity of angiogenic factors.
HRE	Hypoxia-response element. These are short DNA sequences at the promoter level of certain genes (such as VEGF) to which HIF binds to induce expression of target genes.
HUVEC	Human umbilical vein endothelial cells. These cells can be easily isolated from the umbilical cord by treating the internal endothelial surface with enzymes (collagenase) that dissociate endothelial cells of the remaining umbilical cord. Cells are then collected and cultured in culture medium in incubators. HUVECs were used and are used in many angiogenesis and vascular biology studies.
Hypoxia	Decrease of oxygen levels in tissues. Hypoxia is an important stimulator of the expression of certain genes that are involved in cell survival and vascular development.
Integrins	Receptors linked in the plasma membrane that interact with extracellular matrix proteins (fibronectin, laminin, vitronectin, etc.). These are dimeric molecules formed of two units, α and β. The chains α and β can be of different structure ($\alpha 1$, $\alpha 2$, $\alpha 3$, etc.; $\beta 1$, $\beta 2$, $\beta 3$).

Kallicrein	Protein of the blood coagulation. It is part of the endogenous pathway of activation of coagulation more precisely the "contact" system. Kallicrein also has a role in inflammation.
Kernel	A specific set of molecular nodes consisting of molecules with defined relationships.
Leukocytes	Cells of the immune system providing defense against infectious agents. They also play a role in general immunity (also in cancer).
Lipids	Fatty acids. Different types of fatty acids are described (cholesterol, saturated fatty acids, unsaturated fatty acids, etc.).
Lipopolysaccharide (LPS)	Carbohydrate structure molecule (sugar) being coupled to lipid chains (fatty acids). LPS is an agent that induces inflammation and is used in a number of tests.
LNA	Locked nucleic acids through a chemical modification. They are used in molecular diagnostics, and therapeutic development is also underway.
LncRNA (LncRNA)	Longer sequences of non-coding RNA having regulatory functions by an epigenetic mechanism. Their specific roles are only beginning to be understood.
Lumbrokinase	Enzyme of the earthworm that digests blood clots.
Lumen	The vessel lumen consists of a space lined with endothelial cells, space in which the blood circulates.
Lymphatic endothelial cells	Cells specific to the wall of lymphatic vessels in contact with the lymphatic fluid. These cells are characterized by the expression of specific cell surface markers (Podoplanin and Lyve-1).
Lymphatic or venous valves	Mesenchymal structures that prevent reflux of lymphatic fluid or venous blood.
Lymphatics and lymphatic endothelial cells	This is an open absorbent system that originates in the tissues and joins the venous circulation.
Lymphoedema	Lymphoedema is an acquired or inherited condition that is characterized by fluid accumulation in tissues because of rarefaction or malfunction of the lymphatic vessels.
LYVE-1	Lymphatic venous endothelial-1. This is a glycoprotein expressed primarily in lymphatic endothelial cells. It is classically a receptor for hyaluronic acid.
Macromolecules	Large molecules of different chemical natures.
MEM	Minimum essential medium in which certain cells are cultured. DMEM is derived therefrom (see under DMEM).

Metabolome/metabolomics	A large-scale analysis of metabolism and metabolites, such as energy metabolism (breakdown products of glucose metabolism or storage), lipid or amino acids.
Metastasis	Secondary tumor derived from the primary tumor. Tumor cells migrate out from the primary tumor reach the vascular or lymphatic system and colonize organs at a distance.
Methodological and ontological reductionism	Methodological reductionism: simplified methods (experiments, etc.) are used to reduce the complexity of what has to be analyzed. Ontological reductionism: the reduction of a phenomenon to singular events (1 gene for example).
MicroRNA (MIR)	Small, non-coding RNAs that have regulatory functions for the expression of other genes. They can be inhibitory or activating. They are usually named by numbers (MIR-122, 209, etc.).
Netrin	Netrin are guidance molecules of the nervous and vascular system. On axons, these molecules have a repulsive effect. On vessels, the effect is a function of the concentration. Netrin is important in the sprouting and in guidance (directional growth).
Neuropilins	Receptors at the membrane of nerve and endothelial cells. In the nervous system, they are involved in axon guidance. In the vascular system, these molecules serve as co-receptors for VEGF. There are classically two types of neuropilins. Neuropilin-1 (NRP-1) is expressed in the arterial endothelium and Neuropilin-2 (NRP2) in the venous and lymphatic endothelium.
Nucleus (cell nucleus)	Consists of genetic material, chromatin, and associated structural proteins. Surrounded by a nuclear membrane.
Occludin	Intercellular adhesion molecule between two endothelial cells forming part of tight junctions.
Oligodendrocytes	These are glial cells that provide myelin to the central nervous system. These cells can become cancerous and generate oligodendroglyomas.
Oncogenes	Genes whose activation/mutation leads to the formation of cancers. In the normal state, the products of these genes (proteins) have a role in normal cellular functions, especially in cell signaling.
Oxygenation	The supply of oxygen in a tissue.
Peptides	Build of amino acids. Fragment of a protein.

Pericytes	External cells located at the level of the capillaries. They are in close contact with endothelial cells.
Phage display	Phage display is a technology that has been developed to identify new molecules on the surface of cells. There are two approaches: in vivo and in vitro. A phage is an agent that contains genetic material that can infect bacteria and induce their lysis (destruction). However, these phages can also be used as a technological tool. It is thus possible to construct a peptide or peptide-based phage library. These libraries contain multiple peptide sequences or antibody fragments (Each phage contains only one peptide or antibody sequence). After several rounds of selection, it is thus possible to enrich the phage population which contains a specific sequence that selectively binds to a specific cell type or to a specific normal or pathological tissue (tumor for example). This allows then the identification of the receptor to which the specific sequence binds.
Phosphorylation	Introduction of a phosphate molecule into a protein. This is done by kinases. There are two types of kinases: Tyrosine kinases (insertion of the phosphate group at the level of a tyrosine) and serine/threonine kinases (insertion of a phosphate group at the level of a serine or threonine).
Plasma membrane	Membrane formed by a lipid bilayer to which proteins or lipids can be attached.
Plasminogen activators	Enzymes that convert plasminogen to plasmin. Plasmin in turn contributes to the degradation of fibrin ("fibrinolysis") and thus to the dissolution of the clot.
Platelet-derived growth factors (PDGF)	Factors initially purified from blood platelets. Different forms have been identified including PDGF, BB, AA, AB, and CC.
Plathelminth	Flatworm.
PLGF	"Placental growth factor". Growth factor forming part of the VEGF family isolated for the first time in the placenta. There are different forms (PLGF1, 2 ...) and they only bind VEGFR1.
PNA	These are peptide nucleic acids that are similar to DNA or RNA by its bases. They have a different backbone to DNA or RNA consisting of a repetition of *N*-(2-aminoethyl) units)-lycine linked by an amine bond. They have an interest both in diagnostic and therapeutic development.
Progenitor cells	Cells that can give rise to differentiated cells.

Protein C	Protein C is a glycoprotein that has physiological anticoagulant activity. Its synthesis is dependent on vitamin K and it inhibits coagulation factors and favors fibrinolysis.
Protein	Produced after translation from an RNA messenger. Made-up of amino acids.
Proteoglycans	Proteins of the extracellular matrix and proteins inserted into the plasma membrane in contact with the extracellular space comprising heparan sulfate proteoglycans, syndecans, perlecans, etc.
Proteome/proteomics	Study of the expression of proteins on a large scale. This is done by mass spectrometry.
Pruning	Pruning means involution. It is an essential process for maturing the vascular tree.
Receptor-ligand interaction	A factor (usually a protein) that binds to its receptor and activates it.
Receptor	Molecule that can be glycosylated or carry other modifications (phosphorylations etc.) and be bound and activated by a ligand (growth factor, cytokine, hormone, etc.).
Signal transduction—signaling	Sequential activation of molecules that propagate the signal initiated for most of the time by the activation of a receptor. This is because of sequential activation of various substrates by phosphorylation.
Smooth muscle cells	Muscle cells of the vessel wall that have contractile capacity. These cells are distinguished from muscle cells of striated muscles.
Soluble factors	Factor that diffuses in the extracellular medium and in biological fluids (blood, lymphatic fluid, etc.).
Splice variant	During transcription, RNA can be transcribed from different locations. For VEGF-A, this leads to variants of different size and different biological properties. For example, the 121-amino acid VEGF-A weakly binds the matrix whereas 185-amino acid VEGF is a strong matrix binder.
Stabilization of the vessels	When the pericyte coverage of the capillary vessel is incomplete and the interactions between the endothelial cells and the pericytes are not close, the vessel is said destabilized. In this case, the vessel is more sensitive to external factors (stimulators and inhibitors). After stabilization, that is, after establishing close contacts between endothelial cells and pericytes, the vessel has a decreased sensitivity to these factors.

Stem cells	Cells with self-renewal and ability to undergo differentiation. Stem cells are often less sensitive to therapy (resistances).
TAF	"Tumor angiogenesis factor", factor of tumor angiogenesis. It is a common denominator for all factors produced by tumor cells and capable of stimulating angiogenesis.
TGF-β	Transforming growth factor-β. This factor plays an important role in many biological processes such as hematopoiesis, inflammation, tissue repair and cancer. TGF-β is associated with the cell surface to a protein complex called LAP ("Latent activating protein") itself linked to LTBP (Latent TGF-β binding protein). It is activated either by cleavage by plasmin or by thrombospondin-1 which are able to mobilize TEGF-β and remove it from the LAP.
TH lymphocytes	A class of lymphocytes that modulate the immune response.
The cell	Consisting of a nucleus, a plasma membrane, a cytoplasm and internal membrane structures (endoplasmic reticulum, Golgi apparatus, secretory vesicles, etc.).
Thrombospondin	Matricellular protein (present in the matrix produced by cells) of which two forms have been identified. Their effects and mechanisms of action are complex because they inhibit angiogenesis but paradoxically stimulate tumor invasion.
Tight junctions	The tight junctions surround the cell (hence the term "zonula") at the apical pole. They are composed of molecules called occludines and claudines.
TIP and cell stalk cells	Tip cells are located at the end of capillaries and guide the growth of the vascular tube. TIP cells have filipodia that allow their orientation. Stalk cells follow tip cells and form the trunk of the vessel.
TNF	Tumor necrosis factor. There are several types that include TNF-α and TNF-β. These factors have a key role in the regulation of immunity.
tPA	Tissue plasminogen activator (see plasminogen activators).
Transcription factors	Factors that control transcription, that is, the synthesis of messenger RNA from the DNA encoding a specific gene product.

Transcriptome/transcriptomic	Study of gene expression. This consists in the quantification of mRNAs of a cell or tissue sample. Different techniques can be used including PCR, microarrays, and RNA sequencing (RNAseq).
Transglutaminase	Enzyme that connects glutamine. It is involved in the final phase of blood coagulation and is conserved in many species, including invertebrates.
Triploblast	Organism with three embryonic layers.
Tumor initiating cells (CIT)	Stem-like cells with regard to expression markers. These cells are tumorigenic when implanted in animals.
Urochordates	Class of marine animals whose body is surrounded by a tunic. The notochord is only present in the larval state but not in the adult.
Vascular cell map	Mapping of all sequentially occurring cellular interactions occurring in a vascular cell.
Vascular destructive/disruptive agents	Molecules such as Combrestatin that kill endothelial cells. Combrestatin interferes with the cytoskeleton of endothelial cells.
Vascular maturation	After capillary sprouting and growth, newly formed vessels must undergo maturation. This involves the destruction and elimination of vessels in excess. The remaining network allows the establishment of an effective blood flow.
Vascular targeting	Targeting blood vessels by antibodies or chemical molecules. Generally, vessels that express a molecule specifically in a pathological tissue are targeted.
Vascular targeting	This is an alternative strategy to anti-angiogenesis. In the context of targeting, blockade of function is not important. An agent (antibody, chemical molecule) specifically recognizes a molecule (molecular complex) expressed on the surface of newly formed endothelial cells. This is the case in cancer, ischemia, or ocular and inflammatory diseases. It does not recognize preexisting vessels.
Vasculogenesis	Formation of vessels from circulating precursor cells.
VE-cadherin	Intercellular adhesion molecules between two endothelial cells. These undergo homophilic interactions (two VE-cadherin on two different endothelial cells establish contact).
VEGF	Growth factor of the vascular endothelium. This family of factors consists of four subfamilies, called VEGF-A, VEGF-B, VEGF-C, and VEGF-D. VEGFs are considered the predominant vascular growth factors.

Veins	Vessels that bring blood from tissues to the heart.
Weibel–Palade body	Discovered by Weibel and Palade, these organelles have a specific rod-like structure in electron microscopy and contain the von Willebrand factor (factor involved in hemostasis).
Zona occludens (ZO)	See tight junctions.

References

1. Bikfalvi A, Ferrara N (2008) Remembering Jean Plouet, pioneer of angiogenesis research in France and co-discoverer of vascular endothelial growth factor. Cancer Res 68(23)
2. Alitalo K, Tammela T, Petrova TV (2005) Lymphangiogenesis in development and human disease. Nature 438(7070):946–953. https://doi.org/10.1038/nature04480
3. Galen MM (1968) Galen on the usefulness of the parts of the body. Peri chreias morion [romanized form] De usu partium. Cornell University Press, Ithaca, NY
4. Aird WC (2011) Discovery of the cardiovascular system: from Galen to William Harvey. J Thromb Haemostasis 9(Suppl 1):118–129. https://doi.org/10.1111/j.1538-7836.2011.04312.x
5. Keele KPC (1979) Leonardo da Vinci. Corpus of anatomical studies in the collection of her majesty the queen at Windsor Castle. Harcourt Brace Jovanovich, New York
6. Keele K (1972) Leonardo da Vinci's views on arteriosclerosis. XXIII International Congress of the History of Medicine, London, pp 304–308
7. Gharib M, Kremers D, Koochesfahani MM, Kemp M (2002) Leonardo's vision of flow visualization. Exp Fluids 33:219–223
8. Meyerhof M (1935) Ibn al-Nefis (XII century) and his theory of the lesser circulation. Isis 23:100–120
9. Akmal M, Zulkifle M, Ansari AH (2010) IBN Nafis – a forgotten genius in the discovery of pulmonary blood circulation. Heart Views 11:26–30
10. Servetus M (1553) CHRISTIANISMI RESTITUTIO. Restoration of Christianity. An English translation of Christianismi Restitutio. Marian Hillar (Traduction), Christopher A. Hoffman (Traduction) Edwin Mellen Press Ltd, 2006
11. Hofman C, Hillar M (2007/2008) The restauration of Christianity: an English translation of Christianismi restitutio. The Edwin Mellen Press, Lewinston, NY
12. Columbo MR (1559) De re anatomica libri XV. Nicolò Bevilacqua, Venice
13. Carlino A (1999) Books of the body: anatomical ritual and renaissance learning. University of Chicago Press, Chicago
14. Mackall L (1924) A manuscript of the "Christianismi Restitutio" of Servetus, placing the discovery of the pulmonary circulation anterior to 1546. Proc R Soc Med 17(Sect Hist Med):35–38
15. Chereau A (1879) Histoire du Livre, Michael Servet. Bull Acad de Med de Paris 8:30
16. Takrouri MS, Khalaf M (2003) Ibn al-Nafis contributions to science. Middle East J Anaesthesiol 17(2):163–176
17. Cesalpino A (1571) Peripateticarum quaestiorturn lihri quirtqire. Guinta, Venice
18. Fye WB (1996) Andrea Cesalpino. Clin Cardiol 19(12):969–970
19. Giugni F (1959) Further details on the priority in discovery of the circulation of the blood: Andrea Cesalpino or William Harvey? Il Policlinico Sezione pratica 66 (Comunicazioni):1294–1297
20. Aquapendente HF (1603) De Venarum Ostiolis (Trans: Franklin KJ). Published by Charles C Thomas (1933)

A. Bikfalvi, *A Brief History of Blood and Lymphatic Vessels*,
https://doi.org/10.1007/978-3-319-74376-9

21. Scultetus AH, Villavicencio JL, Rich NM (2001) Facts and fiction surrounding the discovery of the venous valves. J Vasc Surg 33(2):435–441. https://doi.org/10.1067/mva.2001.109772
22. Wright T (2012) Circulation: William Harvey's revolutionary idea. Random House, New York
23. Harvey W (1628) Exercitatio anatomica de motu cordis et sanguini in animalibus. Sumptibus Guiliemi Fitzeri, Frankfurt
24. D'Arcy Power FSA (1897) William Harvey. T. FISHER UNWIN, London
25. Franklin KJ (1961) King Charles I and William Harvey. Proc Roy Soc Med 54:85–91
26. Schwann T (1847) Microscopical researches into the accordance in the structure and growth of animals and plants. Sydenham Society, London
27. Malpighi M (1663) De pulmonibus observationibus anatomicæ. De pulmonum substantia & Motu diatribæ. Mathhiæ Godicchii, sumptibus Petri Hauboldi, Copenhagen
28. Bailly M (1771) Nouvelles Litteraires: Vie de Malpighi. In: Observation sur la Physique, sur l'Histoire Naturelle et sur les Arts, Tome Premier, vol Hotel de Thou, Rue des Piotevins. Paris, pp 73–79
29. Van Leeuwenhoek A (1674) Microscopical observations concerning blood, milk, bones, the brain, spittle, and cuticula. Philos Trans 9:121–128
30. Ruysch F (1691) Observationum anatomico-chirurgicarum centuria. Amsterdam 2. Aufl. 1721: 3. Aufl. 1737
31. Earle JW (1835) On the nature of inflammation. Lond Med Gaz 16:6–12
32. Döllinger I (1821) Vom Kreislauf des Blutes. München
33. Dollinger I (1805) Naturlehre des Menschlichen Organismus. Jos. Anton Goebhardt, Bamberg und Würzburg
34. Society R (1831) Dr Marshall Hall on the capillary circulation. Lond Med Gaz 8
35. Müller J (1833) Handbuch der Physiologie des Menschen für Vorlesungen. J. Hölscher Coblenz
36. His W (1865) Die Häute und Höhlen des Körpers. Schwighauser, Basel
37. Koelliker A (1879) Entwicklungsgeschichte des Menschen und der höheren Tiere. Bremen University Press (July 2, 2014), Paperback Edition. Publisher Language: German. ISBN-10:3955626423, Leipzig
38. Müller J, Baly W, Bell J (1843) Elements of physiology. Lea and Blanchard, Philadelphia, PA
39. Müller J (1844) Handbuch der Physiologie des Menschen für Vorlesungen, vol Vierte verbesserte Auflage. Verlag von J. Hölscher, Coblenz
40. Dowler B (1849) Researches, critical and experimental, upon the capillary circulation. New Orleans Med Surg J V:447–477
41. Rouget CBM (1847) Note sur le développement de la tunique contractile des vaisseaux. Comptes rendus hebdomadaires des séances de l' Académie des sciences 79:559–562
42. Zimmermann KW (1923) Der feinere Bau der Blutkapillaren. Z Anat Entwicklungsgesch 68:29–109
43. Mayer S (1902) Die Muskularisierung der capillaren Blutgefässe. Anat Anz 21:442–455
44. Hwa C, Aird WC (2007) The history of the capillary wall: doctors, discoveries, and debates. Am J Phys Heart Circ Phys 293(5):H2667–H2679. https://doi.org/10.1152/ajpheart.00704.2007
45. Slack CB (1843) On abnormal nutrition and on the mode in which its different products are developed. In: Improvements and discoveries in the medical sciences: medical pathology and therapeutics and pratical medicine. Am J Med Sci 441–443
46. Addison W (1843) Experimental and practical researches on inflammation. J. Churchill, London
47. Rather LJ (1972) Addison and the white corpuscles: an aspect of nineteenth-century biology. University of California Press, Berkley, CA
48. Cohnheim J (1867) Ueber Entzündung und Eiterung. Archiv für pathologische Anatomie und Physiologie und für klinische Medicin 40(1–2):1–79

49. Heidland A, Klassen A, Sebekova K, Bahner U (2009) Beginning of modern concept of inflammation: the work of Friedrich Daniel von Recklinghausen and Julius Friedrich Cohnheim. J Nephrol 22(Suppl 14):71–79
50. Clark ER, Clark EI (1935) Observations on changes in blood vascular endothelium in the living animal. Am J Anat 57:388–438
51. Otteley D (1839) John Hunter F.R.S. Haswell, Barrington, and Haswell, Philadelphia, PA
52. Schechter DC, Bergan JJ (1986) Popliteal aneurysm: a celebration of the bicentennial of John Hunter's operation. Ann Vasc Surg 1(1):118–126. https://doi.org/10.1016/S0890-5096(06)60712-7
53. Hunter J (1794) A treatise on the blood, inflammation, and gunshot wounds. Gryphon Editions Ltd., Birmingham
54. Bennett JH (1849) On cancerous and cancroid growths. Prov Med Surg J 13(12):324–326
55. Thiersch C (1869) Der Epithelialkrebs, namentlich der Haut mit Atlas: Eine Anatomisch-Klinische Untersuchung. Verlag von Wilhelm Engelmann, Leipzig
56. Tremain Hertig A (1935) Angiogenesis in the early human chorion and in the primary placenta of the macaque monkey, Contributions to embryology, vol 25. Carnegie Institution of Washington Publication, Washington, DC
57. Clark ER, Hitschler WJ, Kirby-Smith HT, Rex RO, Smith JH (1931) General observations on the ingrowth of new blood vessels into standardized chambers in the rabbit's ear, and the subsequent changes in the newly grown vessels over a period of months. Anat Rec 50:129–167
58. Broca P (1849) Quelques propositions sur les tumeurs dites cancéreuses. Thèse pour le doctorat en Médecine, Paris
59. Broca P (1866) Traité des Tumeurs, vol 1. P Asselin, Paris
60. Goldmann E (1907) The growth of malignant disease in man and the lower animals, with special reference to the vascular system. Lancet 2:1236–1240
61. Goldmann E (1908) The growth of malignant disease in man and the lower animals, with special reference to the vascular system. Proc Roy Soc Med 1(Surg Sect):1–13
62. Greene HS (1941) Heterologous transplantation of mammalian tumors: I. The transfer of rabbit tumors to alien species. J Exp Med 73(4):461–474
63. Greene HS (1941) Heterologous transplantation of mammalian tumors: II. The transfer of human tumors to alien species. J Exp Med 73(4):475–486
64. Greene HS (1938) Heterologous transplantation of human and other mammalian tumors. Science 88(2285):357–358. https://doi.org/10.1126/science.88.2285.357
65. Clark ER, Clark EI (1939) Microscopic observations on the growth of blood capillaries in the living mammals. Am J Anat 64:251–301
66. Algire GH, Chalkley HW (1945) Vascular reactions of normal and malignant tissue in vivo. J Natl Cancer Inst 6:73–85
67. Algire GH, Legallais FY (1948) Growth and vascularization of transplanted mouse melanomas. Ann N Y Acad Sci 4:159–170
68. Merwin RM, Algire GH (1956) The role of graft and host vessels in the vascularization of grafts of normal and neoplastic tissue. J Natl Cancer Inst 17(1):23–33
69. Ide AG, Baker NH, Warren SL (1939) Vascularization of the Brown-Pearce rabbit epithelioma transplant as seen in the transparent ear chambers. AJR 32:891–899
70. Greenblatt M, Shubi P (1968) Tumor angiogenesis: transfilter diffusion studies in the hamster by the transparent chamber technique. J Natl Cancer Inst 41(1):111–124
71. Ehrmann RL, Knoth M (1968) Choriocarcinoma. Transfilter stimulation of vasoproliferation in the hamster cheek pouch. Studied by light and electron microscopy. J Natl Cancer Inst 41 (6):1329–1341
72. Huang X, Molema G, King S, Watkins L, Edgington TS, Thorpe PE (1997) Tumor infarction in mice by antibody-directed targeting of tissue factor to tumor vasculature. Science 275 (5299):547–550

73. Aselli G (1627) De lactibus sive lacteis venis quarto vasorum mesaraicorum genere, novo invento Gaspariis Asellii Cremonensis anatomici Ticinensis dissertatio, qua sententiae anatomicae multae, vel perperam receptae conuelluntur, vel parum perceptae illustrantur. apud Jo. Baptistam Bidellium, Mediolani
74. Pecquet J (1651) Experimenta nova anatomica, quibus incognitum chyli receptaculum, et ab eo per thoracem in ramosusque subclavis vasa lactea deteguntur. Apud Sebastianum Cramoisy et Gabrielem Cramoisy, Paris
75. Bartholin T (1653) Vasa Lymphaticia, Nuper Hafniae in Animantibus Inventa, et Hepatis Exsequiae Holst, Copenhague
76. Rudbeck O (1653) Nova Exercitatio Anatomica Exhibens Ductus Hepaticos Aquosus et Vasa Glandularum Serosa. Lauringer, Vasteras
77. Palmer JF (1837) The works of John Hunter with notes, vol IV. Longman, Rees, Orme, Brown, Green and Longman, London
78. Hewson W (1774) The lymphatic systems in the human subject, and in other animals. J Johnson, London
79. Ludwig C (1852–1856) Lehrbuch der Physiologie des Menschen, vol 2. Akademische Verlagshandlung CF Winter, Heidelberg
80. Fye WB (1986) Carl Ludwig and the Leipzig Physiological Institute: 'a factory of new knowledge'. Circulation 74(5):920–928
81. von Recklinghausen F (1860) Eine methode, mikroskopische hohle und solide gebilde voneinander zu unterscheiden. Virchows Arch 19
82. Sabin FR (1916) The method of growth of the lymphatic system. Science 44(1127):145–158. https://doi.org/10.1126/science.44.1127.145
83. Sabin FR (1902) On the origin of the lymphatic system from the veins and the development of the lymph hearts and thoracic duct in the pig. Am J Anat 1:367–389
84. Clark ER (1909) Observations on living growing lymphatics in the tail of the frog larva. Anat Rec 6(261)
85. Clark ER, Clark EL (1933) Further observations on living lymphatic vessels in the transparent chamber in the rabbit's ear-their relation to the tissue spaces. Am J Anat 52:273
86. Clark ER, Clark EL (1937) Observations on living mammalian lymphatic capillaries-their relation to the blood vessels. Am J Anat 60:253
87. Clark ER, Clark EL (1938) Observations of isolated lymphatic capillaries in the living mammal. Am J Anat 62(1):59–92
88. Monahan-Earley R, Dvorak AM, Aird WC (2013) Evolutionary origins of the blood vascular system and endothelium. J Thromb Haemost 11(Suppl 1):46–66. https://doi.org/10.1111/jth.12253
89. Cerenius L, Soderhall K (2011) Coagulation in invertebrates. J Innate Immunity 3(1):3–8. https://doi.org/10.1159/000322066
90. Maningas MB, Kondo H, Hirono I (2013) Molecular mechanisms of the shrimp clotting system. Fish Shellfish Immunol 34(4):968–972. https://doi.org/10.1016/j.fsi.2012.09.018
91. Theopold U, Krautz R, Dushay MS (2014) The Drosophila clotting system and its messages for mammals. Dev Comp Immunol 42(1):42–46. https://doi.org/10.1016/j.dci.2013.03.014
92. Cho NK, Keyes L, Johnson E, Heller J, Ryner L, Karim F, Krasnow MA (2002) Developmental control of blood cell migration by the Drosophila VEGF pathway. Cell 108(6):865–876
93. Parsons B, Foley E (2013) The Drosophila platelet-derived growth factor and vascular endothelial growth factor-receptor related (Pvr) protein ligands Pvf2 and Pvf3 control hemocyte viability and invasive migration. J Biol Chem 288(28):20173–20183. https://doi.org/10.1074/jbc.M113.483818
94. Fandrey J, Gorr TA, Gassmann M (2006) Regulating cellular oxygen sensing by hydroxylation. Cardiovasc Res 71(4):642–651. https://doi.org/10.1016/j.cardiores.2006.05.005
95. Gorr TA, Gassmann M, Wappner P (2006) Sensing and responding to hypoxia via HIF in model invertebrates. J Insect Physiol 52(4):349–364. https://doi.org/10.1016/j.jinsphys.2006.01.002

96. Tiozzo S, Voskoboynik A, Brown FD, De Tomaso AW (2008) A conserved role of the VEGF pathway in angiogenesis of an ectodermally-derived vasculature. Dev Biol 315(1):243–255. https://doi.org/10.1016/j.ydbio.2007.12.035
97. Gasparini F, Caicci F, Rigon F, Zaniolo G, Manni L (2014) Testing an unusual in vivo vessel network model: a method to study angiogenesis in the colonial tunicate Botryllus schlosseri. Sci Rep 4:6460. https://doi.org/10.1038/srep06460
98. Algire GH, Chalkley HW, Earle WE, Legallais FY, Park HD, Shelton E, Schilling EL (1950) Vascular reactions of normal and malignant tissues in vivo. III Vascular reactions' of mice to fibroblasts treated in vitro with methylcholanthrene. J Natl Cancer Inst 11(3):555–580
99. Michaelson IC (1948) The mode of development of the vascular system of the retina, with some observations on its significance for certain retinal disease. Trans Ophthalmol Soc UK 68:137–180
100. Campbell FM (1951) The influence of a low atmospheric pressure on the development of the retinal vessels in the rat. Trans Ophthalmol Soc UK 71:287–300
101. Gimbrone MA Jr, Leapman SB, Cotran RS, Folkman J (1972) Tumor dormancy in vivo by prevention of neovascularization. J Exp Med 136(2):261–276
102. Folkman J, Merler E, Abernathy C, Williams G (1971) Isolation of a tumor factor responsible for angiogenesis. J Exp Med 133(2):275–288
103. Folkman J (1971) Transplacental carcinogenesis by stilbestrol. N Engl J Med 285(7):404–405. https://doi.org/10.1056/NEJM197108122850711
104. Harrison R (1907) Observations on the living developing nerve fiber. Anat Rec, Proc Soc Exp Med, NY 1:116–128
105. Carrel A (1912) Pure cultures of cells. J Exp Med 16(2):165–168
106. Carrel A (1912) On the permanent life of tissues outside of the organism. J Exp Med 15 (5):516–528
107. Eagle H (1955) Nutrition needs of mammalian cells in tissue culture. Science 122 (3168):501–514
108. Jaffe EA, Nachman RL, Becker CG, Minick CR (1973) Culture of human endothelial cells derived from umbilical veins. Identification by morphologic and immunologic criteria. J Clin Invest 52(11):2745–2756. https://doi.org/10.1172/JCI107470
109. Gimbrone MA Jr, Cotran RS, Folkman J (1973) Endothelial regeneration: studies with human endothelial cells in culture. Ser Haematol 6(4):453–455
110. Campbell GR, Uehara Y, Mark G, Burnstock G (1971) Fine structure of smooth muscle cells grown in tissue culture. J Cell Biol 49(1):21–34
111. Chamley-Campbell J, Campbell GR, Ross R (1979) The smooth muscle cell in culture. Physiol Rev 59(1):1–61
112. Buzney SM, Frank RN, Robison WG Jr (1975) Retinal capillaries: proliferation of mural cells in vitro. Science 190(4218):985–986
113. Weibel ER, Palade GE (1964) New cytoplasmic components in arterial endothelia. J Cell Biol 23:101–112
114. Shing Y, Folkman J, Sullivan R, Butterfield C, Murray J, Klagsbrun M (1984) Heparin affinity: purification of a tumor-derived capillary endothelial cell growth factor. Science 223 (4612):1296–1299
115. Klagsbrun M, Sullivan R, Smith S, Rybka R, Shing YE (1987) Purification of endothelial cell growth factors by heparin affinity chromatography. Methods Enzymol 147:95–105
116. Lobb R, Sasse J, Sullivan R, Shing Y, D'Amore P, Jacobs J, Klagsbrun M (1986) Purification and characterization of heparin-binding endothelial cell growth factors. J Biol Chem 261 (4):1924–1928
117. Schreiber AB, Kenney J, Kowalski J, Thomas KA, Gimenez-Gallego G, Rios-Candelore M, Di Salvo J, Barritault D, Courty J, Courtois Y, Moenner M, Loret C, Burgess WH, Mehlman T, Friesel R, Johnson W, Maciag T (1985) A unique family of endothelial cell polypeptide mitogens: the antigenic and receptor cross-reactivity of bovine endothelial cell

growth factor, brain-derived acidic fibroblast growth factor, and eye-derived growth factor-II. J Cell Biol 101(4):1623–1626
118. Risau W (1996) What, if anything, is an angiogenic factor? Cancer Metastasis Rev 15 (2):149–151
119. Fett JW, Strydom DJ, Lobb RR, Alderman EM, Bethune JL, Riordan JF, Vallee BL (1985) Isolation and characterization of angiogenin, an angiogenic protein from human carcinoma cells. Biochemistry 24(20):5480–5486
120. Strydom DJ, Fett JW, Lobb RR, Alderman EM, Bethune JL, Riordan JF, Vallee BL (1985) Amino acid sequence of human tumor derived angiogenin. Biochemistry 24(20):5486–5494
121. Klagsbrun M, Shing Y (1985) Heparin affinity of anionic and cationic capillary endothelial cell growth factors: analysis of hypothalamus-derived growth factors and fibroblast growth factors. Proc Natl Acad Sci U S A 82(3):805–809
122. Plouet J, Schilling J, Gospodarowicz D (1989) Isolation and characterization of a newly identified endothelial cell mitogen produced by AtT-20 cells. EMBO J 8(12):3801–3806
123. Leung DW, Cachianes G, Kuang WJ, Goeddel DV, Ferrara N (1989) Vascular endothelial growth factor is a secreted angiogenic mitogen. Science 246(4935):1306–1309
124. Senger DR, Perruzzi CA, Feder J, Dvorak HF (1986) A highly conserved vascular permeability factor secreted by a variety of human and rodent tumor cell lines. Cancer Res 46 (11):5629–5632
125. de Vries C, Escobedo JA, Ueno H, Houck K, Ferrara N, Williams LT (1992) The fms-like tyrosine kinase, a receptor for vascular endothelial growth factor. Science 255(5047):989–991
126. Terman BI, Dougher-Vermazen M, Carrion ME, Dimitrov D, Armellino DC, Gospodarowicz D, Bohlen P (1992) Identification of the KDR tyrosine kinase as a receptor for vascular endothelial cell growth factor. Biochem Biophys Res Commun 187(3):1579–1586
127. Makinen T, Veikkola T, Mustjoki S, Karpanen T, Catimel B, Nice EC, Wise L, Mercer A, Kowalski H, Kerjaschki D, Stacker SA, Achen MG, Alitalo K (2001) Isolated lymphatic endothelial cells transduce growth, survival and migratory signals via the VEGF-C/D receptor VEGFR-3. EMBO J 20(17):4762–4773. https://doi.org/10.1093/emboj/20.17.4762
128. Waltenberger J, Claesson-Welsh L, Siegbahn A, Shibuya M, Heldin CH (1994) Different signal transduction properties of KDR and Flt1, two receptors for vascular endothelial growth factor. J Biol Chem 269(43):26988–26995
129. Carmeliet P, Ferreira V, Breier G, Pollefeyt S, Kieckens L, Gertsenstein M, Fahrig M, Vandenhoeck A, Harpal K, Eberhardt C, Declercq C, Pawling J, Moons L, Collen D, Risau W, Nagy A (1996) Abnormal blood vessel development and lethality in embryos lacking a single VEGF allele. Nature 380(6573):435–439. https://doi.org/10.1038/380435a0
130. Ferrara N, Carver-Moore K, Chen H, Dowd M, Lu L, O'Shea KS, Powell-Braxton L, Hillan KJ, Moore MW (1996) Heterozygous embryonic lethality induced by targeted inactivation of the VEGF gene. Nature 380(6573):439–442. https://doi.org/10.1038/380439a0
131. Kim KJ, Li B, Winer J, Armanini M, Gillett N, Phillips HS, Ferrara N (1993) Inhibition of vascular endothelial growth factor-induced angiogenesis suppresses tumour growth in vivo. Nature 362(6423):841–844. https://doi.org/10.1038/362841a0
132. Folkman J (2006) Angiogenesis. Annu Rev Med 57:1–18. https://doi.org/10.1146/annurev.med.57.121304.131306
133. O'Reilly MS, Boehm T, Shing Y, Fukai N, Vasios G, Lane WS, Flynn E, Birkhead JR, Olsen BR, Folkman J (1997) Endostatin: an endogenous inhibitor of angiogenesis and tumor growth. Cell 88(2):277–285
134. Boehm T, Folkman J, Browder T, O'Reilly MS (1997) Antiangiogenic therapy of experimental cancer does not induce acquired drug resistance. Nature 390(6658):404–407. https://doi.org/10.1038/37126
135. Kolata G (1998) Two drugs eradicate tumors in mice. New York Times, May 3
136. Xie L, Duncan MB, Pahler J, Sugimoto H, Martino M, Lively J, Mundel T, Soubasakos M, Rubin K, Takeda T, Inoue M, Lawler J, Hynes RO, Hanahan D, Kalluri R (2011)

Counterbalancing angiogenic regulatory factors control the rate of cancer progression and survival in a stage-specific manner. Proc Natl Acad Sci U S A 108(24):9939–9944. https://doi.org/10.1073/pnas.1105041108
137. Murphy-Ullrich JE, Sage EH (2014) Revisiting the matricellular concept. J Int Soc Matrix Biol 37:1–14. https://doi.org/10.1016/j.matbio.2014.07.005
138. Ferrara N, Hillan KJ, Novotny W (2005) Bevacizumab (Avastin), a humanized anti-VEGF monoclonal antibody for cancer therapy. Biochem Biophys Res Commun 333(2):328–335. https://doi.org/10.1016/j.bbrc.2005.05.132
139. Hurwitz H, Fehrenbacher L, Novotny W, Cartwright T, Hainsworth J, Heim W, Berlin J, Baron A, Griffing S, Holmgren E, Ferrara N, Fyfe G, Rogers B, Ross R, Kabbinavar F (2004) Bevacizumab plus irinotecan, fluorouracil, and leucovorin for metastatic colorectal cancer. N Engl J Med 350(23):2335–2342. https://doi.org/10.1056/NEJMoa032691
140. Yancopoulos GD (2010) Clinical application of therapies targeting VEGF. Cell 143(1):13–16. https://doi.org/10.1016/j.cell.2010.09.028
141. Berger JS, Hiatt WR (2012) Medical therapy in peripheral artery disease. Circulation 126 (4):491–500. https://doi.org/10.1161/CIRCULATIONAHA.111.033886
142. Comerota AJ, Throm RC, Miller KA, Henry T, Chronos N, Laird J, Sequeira R, Kent CK, Bacchetta M, Goldman C, Salenius JP, Schmieder FA, Pilsudski R (2002) Naked plasmid DNA encoding fibroblast growth factor type 1 for the treatment of end-stage unreconstructible lower extremity ischemia: preliminary results of a phase I trial. J Vasc Surg 35(5):930–936
143. Lederman RJ, Mendelsohn FO, Anderson RD, Saucedo JF, Tenaglia AN, Hermiller JB, Hillegass WB, Rocha-Singh K, Moon TE, Whitehouse MJ, Annex BH, Investigators T (2002) Therapeutic angiogenesis with recombinant fibroblast growth factor-2 for intermittent claudication (the TRAFFIC study): a randomised trial. Lancet 359(9323):2053–2058
144. Khurana R, Simons M (2003) Insights from angiogenesis trials using fibroblast growth factor for advanced arteriosclerotic disease. Trends Cardiovasc Med 13(3):116–122
145. Fadini GP, Losordo D, Dimmeler S (2012) Critical reevaluation of endothelial progenitor cell phenotypes for therapeutic and diagnostic use. Circ Res 110(4):624–637. https://doi.org/10.1161/CIRCRESAHA.111.243386
146. Rohde E, Schallmoser K, Reinisch A, Hofmann NA, Pfeifer T, Frohlich E, Rechberger G, Lanzer G, Kratky D, Strunk D (2011) Pro-angiogenic induction of myeloid cells for therapeutic angiogenesis can induce mitogen-activated protein kinase p38-dependent foam cell formation. Cytotherapy 13(4):503–512. https://doi.org/10.3109/14653249.2010.536214
147. Gerhardt H, Golding M, Fruttiger M, Ruhrberg C, Lundkvist A, Abramsson A, Jeltsch M, Mitchell C, Alitalo K, Shima D, Betsholtz C (2003) VEGF guides angiogenic sprouting utilizing endothelial tip cell filopodia. J Cell Biol 161(6):1163–1177. https://doi.org/10.1083/jcb.200302047
148. Gerhardt H, Betsholtz C (2005) How do endothelial cells orientate? EXS 94:3–15
149. Benedito R, Rocha SF, Woeste M, Zamykal M, Radtke F, Casanovas O, Duarte A, Pytowski B, Adams RH (2012) Notch-dependent VEGFR3 upregulation allows angiogenesis without VEGF-VEGFR2 signalling. Nature 484(7392):110–114. https://doi.org/10.1038/nature10908
150. Zarkada G, Heinolainen K, Makinen T, Kubota Y, Alitalo K (2015) VEGFR3 does not sustain retinal angiogenesis without VEGFR2. Proc Natl Acad Sci U S A 112(3):761–766. https://doi.org/10.1073/pnas.1423278112
151. Tamariz E, Varela-Echavarria A (2015) The discovery of the growth cone and its influence on the study of axon guidance. Front Neuroanat 9:51. https://doi.org/10.3389/fnana.2015.00051
152. Pelton JC, Wright CE, Leitges M, Bautch VL (2014) Multiple endothelial cells constitute the tip of developing blood vessels and polarize to promote lumen formation. Development 141 (21):4121–4126. https://doi.org/10.1242/dev.110296
153. Thomas JL, Baker K, Han J, Calvo C, Nurmi H, Eichmann AC, Alitalo K (2013) Interactions between VEGFR and Notch signaling pathways in endothelial and neural cells. Cell Mol Life Sci 70(10):1779–1792. https://doi.org/10.1007/s00018-013-1312-6

154. Ochsenbein AM, Karaman S, Proulx ST, Berchtold M, Jurisic G, Stoeckli ET, Detmar M (2016) Endothelial cell-derived semaphorin 3A inhibits filopodia formation by blood vascular tip cells. Development 143(4):589–594. https://doi.org/10.1242/dev.127670
155. Teuwen LA, Draoui N, Dubois C, Carmeliet P (2017) Endothelial cell metabolism: an update anno 2017. Curr Opin Hematol. https://doi.org/10.1097/MOH.0000000000000335
156. Kur E, Kim J, Tata A, Comin CH, Harrington KI, Costa L da F, Bentley K, Gu C (2016) Temporal modulation of collective cell behavior controls vascular network topology. eLife 5. https://doi.org/10.7554/eLife.13212
157. Sigurbjornsdottir S, Mathew R, Leptin M (2014) Molecular mechanisms of de novo lumen formation. Nat Rev Mol Cell Biol 15(10):665–676. https://doi.org/10.1038/nrm3871
158. Strilic B, Kucera T, Eglinger J, Hughes MR, McNagny KM, Tsukita S, Dejana E, Ferrara N, Lammert E (2009) The molecular basis of vascular lumen formation in the developing mouse aorta. Dev Cell 17(4):505–515. https://doi.org/10.1016/j.devcel.2009.08.011
159. Kucera T, Strilic B, Regener K, Schubert M, Laudet V, Lammert E (2009) Ancestral vascular lumen formation via basal cell surfaces. PLoS One 4(1):e4132. https://doi.org/10.1371/journal.pone.0004132
160. Kamei M, Saunders WB, Bayless KJ, Dye L, Davis GE, Weinstein BM (2006) Endothelial tubes assemble from intracellular vacuoles in vivo. Nature 442(7101):453–456. https://doi.org/10.1038/nature04923
161. Herwig L, Blum Y, Krudewig A, Ellertsdottir E, Lenard A, Belting HG, Affolter M (2011) Distinct cellular mechanisms of blood vessel fusion in the zebrafish embryo. Curr Biol 21 (22):1942–1948. https://doi.org/10.1016/j.cub.2011.10.016
162. Axnick J, Lammert E (2012) Vascular lumen formation. Curr Opin Hematol 19(3):192–198. https://doi.org/10.1097/MOH.0b013e3283523ebc
163. Phng LK, Gebala V, Bentley K, Philippides A, Wacker A, Mathivet T, Sauteur L, Stanchi F, Belting HG, Affolter M, Gerhardt H (2015) Formin-mediated actin polymerization at endothelial junctions is required for vessel lumen formation and stabilization. Dev Cell 32 (1):123–132. https://doi.org/10.1016/j.devcel.2014.11.017
164. Gebala V, Collins R, Geudens I, Phng LK, Gerhardt H (2016) Blood flow drives lumen formation by inverse membrane blebbing during angiogenesis in vivo. Nat Cell Biol 18 (4):443–450. https://doi.org/10.1038/ncb3320
165. Yang Y, Oliver G (2014) Development of the mammalian lymphatic vasculature. J Clin Invest 124(3):888–897. https://doi.org/10.1172/JCI71609
166. Sabine A, Agalarov Y, Maby-El Hajjami H, Jaquet M, Hagerling R, Pollmann C, Bebber D, Pfenniger A, Miura N, Dormond O, Calmes JM, Adams RH, Makinen T, Kiefer F, Kwak BR, Petrova TV (2012) Mechanotransduction, PROX1, and FOXC2 cooperate to control connexin37 and calcineurin during lymphatic-valve formation. Dev Cell 22(2):430–445. https://doi.org/10.1016/j.devcel.2011.12.020
167. Trani M, Dejana E (2015) New insights in the control of vascular permeability: vascular endothelial-cadherin and other players. Curr Opin Hematol. https://doi.org/10.1097/MOH.0000000000000137
168. Le Guelte A, Dwyer J, Gavard J (2011) Jumping the barrier: VE-cadherin, VEGF and other angiogenic modifiers in cancer. Biol Cell 103(12):593–605. https://doi.org/10.1042/BC20110069
169. Azzi S, Hebda JK, Gavard J (2013) Vascular permeability and drug delivery in cancers. Front Oncol 3:211. https://doi.org/10.3389/fonc.2013.00211
170. Joukov V, Pajusola K, Kaipainen A, Chilov D, Lahtinen I, Kukk E, Saksela O, Kalkkinen N, Alitalo K (1996) A novel vascular endothelial growth factor, VEGF-C, is a ligand for the Flt4 (VEGFR-3) and KDR (VEGFR-2) receptor tyrosine kinases. EMBO J 15(7):1751
171. Siegfried G, Basak A, Cromlish JA, Benjannet S, Marcinkiewicz J, Chretien M, Seidah NG, Khatib AM (2003) The secretory proprotein convertases furin, PC5, and PC7 activate VEGF-C to induce tumorigenesis. J Clin Invest 111(11):1723–1732. https://doi.org/10.1172/JCI17220

172. Fischer C, Jonckx B, Mazzone M, Zacchigna S, Loges S, Pattarini L, Chorianopoulos E, Liesenborghs L, Koch M, De Mol M, Autiero M, Wyns S, Plaisance S, Moons L, van Rooijen N, Giacca M, Stassen JM, Dewerchin M, Collen D, Carmeliet P (2007) Anti-PlGF inhibits growth of VEGF(R)-inhibitor-resistant tumors without affecting healthy vessels. Cell 131(3):463–475. https://doi.org/10.1016/j.cell.2007.08.038
173. Dewerchin M, Carmeliet P (2014) Placental growth factor in cancer. Expert Opin Ther Targets 18(11):1339–1354. https://doi.org/10.1517/14728222.2014.948420
174. Eriksson A, Cao R, Pawliuk R, Berg SM, Tsang M, Zhou D, Fleet C, Tritsaris K, Dissing S, Leboulch P, Cao Y (2002) Placenta growth factor-1 antagonizes VEGF-induced angiogenesis and tumor growth by the formation of functionally inactive PlGF-1/VEGF heterodimers. Cancer Cell 1(1):99–108
175. Yang X, Zhang Y, Yang Y, Lim S, Cao Z, Rak J, Cao Y (2013) Vascular endothelial growth factor-dependent spatiotemporal dual roles of placental growth factor in modulation of angiogenesis and tumor growth. Proc Natl Acad Sci U S A 110(34):13932–13937. https://doi.org/10.1073/pnas.1309629110
176. Davis S, Yancopoulos GD (1999) The angiopoietins: Yin and Yang in angiogenesis. Curr Top Microbiol Immunol 237:173–185
177. Koh GY (2013) Orchestral actions of angiopoietin-1 in vascular regeneration. Trends Mol Med 19(1):31–39. https://doi.org/10.1016/j.molmed.2012.10.010
178. Auguste P, Gursel DB, Lemiere S, Reimers D, Cuevas P, Carceller F, Di Santo JP, Bikfalvi A (2001) Inhibition of fibroblast growth factor/fibroblast growth factor receptor activity in glioma cells impedes tumor growth by both angiogenesis-dependent and -independent mechanisms. Cancer Res 61(4):1717–1726
179. Larrieu-Lahargue F, Welm AL, Bouchecareilh M, Alitalo K, Li DY, Bikfalvi A, Auguste P (2012) Blocking Fibroblast Growth Factor receptor signaling inhibits tumor growth, lymphangiogenesis, and metastasis. PLoS One 7(6):e39540. https://doi.org/10.1371/journal.pone.0039540
180. Shin JW, Min M, Larrieu-Lahargue F, Canron X, Kunstfeld R, Nguyen L, Henderson JE, Bikfalvi A, Detmar M, Hong YK (2006) Prox1 promotes lineage-specific expression of fibroblast growth factor (FGF) receptor-3 in lymphatic endothelium: a role for FGF signaling in lymphangiogenesis. Mol Biol Cell 17(2):576–584. https://doi.org/10.1091/mbc.E05-04-0368
181. Rousseau B, Larrieu-Lahargue F, Javerzat S, Guilhem-Ducleon F, Beermann F, Bikfalvi A (2004) The tyrp1-Tag/tyrp1-FGFR1-DN bigenic mouse: a model for selective inhibition of tumor development, angiogenesis, and invasion into the neural tissue by blockade of fibroblast growth factor receptor activity. Cancer Res 64(7):2490–2495
182. Casanovas O, Hicklin DJ, Bergers G, Hanahan D (2005) Drug resistance by evasion of antiangiogenic targeting of VEGF signaling in late-stage pancreatic islet tumors. Cancer Cell 8(4):299–309. https://doi.org/10.1016/j.ccr.2005.09.005
183. Murakami M, Nguyen LT, Zhuang ZW, Moodie KL, Carmeliet P, Stan RV, Simons M (2008) The FGF system has a key role in regulating vascular integrity. J Clin Invest 118 (10):3355–3366. https://doi.org/10.1172/JCI35298
184. De Smet F, Tembuyser B, Lenard A, Claes F, Zhang J, Michielsen C, Van Schepdael A, Herbert JM, Bono F, Affolter M, Dewerchin M, Carmeliet P (2014) Fibroblast growth factor signaling affects vascular outgrowth and is required for the maintenance of blood vessel integrity. Chem Biol 21(10):1310–1317. https://doi.org/10.1016/j.chembiol.2014.07.018
185. Yu P, Wilhelm K, Dubrac A, Tung JK, Alves TC, Fang JS, Xie Y, Zhu J, Chen Z, De Smet F, Zhang J, Jin SW, Sun L, Sun H, Kibbey RG, Hirschi KK, Hay N, Carmeliet P, Chittenden TW, Eichmann A, Potente M, Simons M (2017) FGF-dependent metabolic control of vascular development. Nature 545(7653):224–228. https://doi.org/10.1038/nature22322
186. Gerhardt H, Betsholtz C (2003) Endothelial-pericyte interactions in angiogenesis. Cell Tissue Res 314(1):15–23. https://doi.org/10.1007/s00441-003-0745-x
187. Lindblom P, Gerhardt H, Liebner S, Abramsson A, Enge M, Hellstrom M, Backstrom G, Fredriksson S, Landegren U, Nystrom HC, Bergstrom G, Dejana E, Ostman A, Lindahl P, Betsholtz C (2003) Endothelial PDGF-B retention is required for proper investment of

pericytes in the microvessel wall. Genes Dev 17(15):1835–1840. https://doi.org/10.1101/gad.266803
188. Gaengel K, Genove G, Armulik A, Betsholtz C (2009) Endothelial-mural cell signaling in vascular development and angiogenesis. Arterioscler Thromb Vasc Biol 29(5):630–638. https://doi.org/10.1161/ATVBAHA.107.161521
189. Ferrara N (2010) Role of myeloid cells in vascular endothelial growth factor-independent tumor angiogenesis. Curr Opin Hematol 17(3):219–224. https://doi.org/10.1097/MOH.0b013e3283386660
190. Chung AS, Wu X, Zhuang G, Ngu H, Kasman I, Zhang J, Vernes JM, Jiang Z, Meng YG, Peale FV, Ouyang W, Ferrara N (2013) An interleukin-17-mediated paracrine network promotes tumor resistance to anti-angiogenic therapy. Nat Med 19(9):1114–1123. https://doi.org/10.1038/nm.3291
191. Lobov IB, Rao S, Carroll TJ, Vallance JE, Ito M, Ondr JK, Kurup S, Glass DA, Patel MS, Shu W, Morrisey EE, McMahon AP, Karsenty G, Lang RA (2005) WNT7b mediates macrophage-induced programmed cell death in patterning of the vasculature. Nature 437 (7057):417–421. https://doi.org/10.1038/nature03928
192. Stefater JA III, Lewkowich I, Rao S, Mariggi G, Carpenter AC, Burr AR, Fan J, Ajima R, Molkentin JD, Williams BO, Wills-Karp M, Pollard JW, Yamaguchi T, Ferrara N, Gerhardt H, Lang RA (2011) Regulation of angiogenesis by a non-canonical Wnt-Flt1 pathway in myeloid cells. Nature 474(7352):511–515. https://doi.org/10.1038/nature10085
193. Wynn TA, Chawla A, Pollard JW (2013) Macrophage biology in development, homeostasis and disease. Nature 496(7446):445–455. https://doi.org/10.1038/nature12034
194. Jetten N, Verbruggen S, Gijbels MJ, Post MJ, De Winther MP, Donners MM (2014) Anti-inflammatory M2, but not pro-inflammatory M1 macrophages promote angiogenesis in vivo. Angiogenesis 17(1):109–118. https://doi.org/10.1007/s10456-013-9381-6
195. Gomez Perdiguero E, Klapproth K, Schulz C, Busch K, Azzoni E, Crozet L, Garner H, Trouillet C, de Bruijn MF, Geissmann F, Rodewald HR (2015) Tissue-resident macrophages originate from yolk-sac-derived erythro-myeloid progenitors. Nature 518(7540):547–551. https://doi.org/10.1038/nature13989
196. Rymo SF, Gerhardt H, Wolfhagen Sand F, Lang R, Uv A, Betsholtz C (2011) A two-way communication between microglial cells and angiogenic sprouts regulates angiogenesis in aortic ring cultures. PLoS One 6(1):e15846. https://doi.org/10.1371/journal.pone.0015846
197. Zhang F, Li Y, Tang Z, Kumar A, Lee C, Zhang L, Zhu C, Klotzsche-von Ameln A, Wang B, Gao Z, Zhang S, Langer HF, Hou X, Jensen L, Ma W, Wong W, Chavakis T, Liu Y, Cao Y, Li X (2012) Proliferative and survival effects of PUMA promote angiogenesis. Cell Rep 2 (5):1272–1285. https://doi.org/10.1016/j.celrep.2012.09.023
198. Hanahan D, Folkman J (1996) Patterns and emerging mechanisms of the angiogenic switch during tumorigenesis. Cell 86(3):353–364
199. Semenza GL (2012) Hypoxia-inducible factors: mediators of cancer progression and targets for cancer therapy. Trends Pharmacol Sci 33(4):207–214. https://doi.org/10.1016/j.tips.2012.01.005
200. Kaelin WG Jr, Ratcliffe PJ (2008) Oxygen sensing by metazoans: the central role of the HIF hydroxylase pathway. Mol Cell 30(4):393–402. https://doi.org/10.1016/j.molcel.2008.04.009
201. Berra E, Pages G, Pouyssegur J (2000) MAP kinases and hypoxia in the control of VEGF expression. Cancer Metastasis Rev 19(1–2):139–145
202. Dews M, Homayouni A, Yu D, Murphy D, Sevignani C, Wentzel E, Furth EE, Lee WM, Enders GH, Mendell JT, Thomas-Tikhonenko A (2006) Augmentation of tumor angiogenesis by a Myc-activated microRNA cluster. Nat Genet 38(9):1060–1065. https://doi.org/10.1038/ng1855
203. Shibuya M (2013) VEGFR and type-V RTK activation and signaling. Cold Spring Harb Perspect Biol 5(10):a009092. https://doi.org/10.1101/cshperspect.a009092
204. Lampropoulou A, Ruhrberg C (2014) Neuropilin regulation of angiogenesis. Biochem Soc Trans 42(6):1623–1628. https://doi.org/10.1042/BST20140244

205. Murakami M, Simons M (2008) Fibroblast growth factor regulation of neovascularization. Curr Opin Hematol 15(3):215–220. https://doi.org/10.1097/MOH.0b013e3282f97d98
206. Hawinkels LJ, Garcia de Vinuesa A, Ten Dijke P (2013) Activin receptor-like kinase 1 as a target for anti-angiogenesis therapy. Expert Opin Investig Drugs 22(11):1371–1383. https://doi.org/10.1517/13543784.2013.837884
207. Freitas C, Larrivee B, Eichmann A (2008) Netrins and UNC5 receptors in angiogenesis. Angiogenesis 11(1):23–29. https://doi.org/10.1007/s10456-008-9096-2
208. Warburg O (1956) On the origin of cancer cells. Science 123(3191):309–314
209. Warburg O (1956) On respiratory impairment in cancer cells. Science 124(3215):269–270
210. Verdegem D, Moens S, Stapor P, Carmeliet P (2014) Endothelial cell metabolism: parallels and divergences with cancer cell metabolism. Cancer Metab 2:19. https://doi.org/10.1186/2049-3002-2-19
211. Schoors S, Bruning U, Missiaen R, Queiroz KC, Borgers G, Elia I, Zecchin A, Cantelmo AR, Christen S, Goveia J, Heggermont W, Godde L, Vinckier S, Van Veldhoven PP, Eelen G, Schoonjans L, Gerhardt H, Dewerchin M, Baes M, De Bock K, Ghesquiere B, Lunt SY, Fendt SM, Carmeliet P (2015) Fatty acid carbon is essential for dNTP synthesis in endothelial cells. Nature 520(7546):192–197. https://doi.org/10.1038/nature14362
212. Lazarus A, Del-Moral PM, Ilovich O, Mishani E, Warburton D, Keshet E (2011) A perfusion-independent role of blood vessels in determining branching stereotypy of lung airways. Development 138(11):2359–2368. https://doi.org/10.1242/dev.060723
213. Magenheim J, Ilovich O, Lazarus A, Klochendler A, Ziv O, Werman R, Hija A, Cleaver O, Mishani E, Keshet E, Dor Y (2011) Blood vessels restrain pancreas branching, differentiation and growth. Development 138(21):4743–4752. https://doi.org/10.1242/dev.066548
214. Lammert E, Cleaver O, Melton D (2003) Role of endothelial cells in early pancreas and liver development. Mech Dev 120(1):59–64
215. Lammert E, Gu G, McLaughlin M, Brown D, Brekken R, Murtaugh LC, Gerber HP, Ferrara N, Melton DA (2003) Role of VEGF-A in vascularization of pancreatic islets. Curr Biol 13(12):1070–1074
216. Rafii S, Butler JM, Ding BS (2016) Angiocrine functions of organ-specific endothelial cells. Nature 529(7586):316–325. https://doi.org/10.1038/nature17040
217. Kim J, Kang Y, Kojima Y, Lighthouse JK, Hu X, Aldred MA, McLean DL, Park H, Comhair SA, Greif DM, Erzurum SC, Chun HJ (2013) An endothelial apelin-FGF link mediated by miR-424 and miR-503 is disrupted in pulmonary arterial hypertension. Nat Med 19(1):74–82. https://doi.org/10.1038/nm.3040
218. Yang P, Read C, Kuc RE, Buonincontri G, Southwood M, Torella R, Upton PD, Crosby A, Sawiak SJ, Carpenter TA, Glen RC, Morrell NW, Maguire JJ, Davenport AP (2017) Elabela/toddler is an endogenous agonist of the apelin APJ receptor in the adult cardiovascular system, and exogenous administration of the peptide compensates for the downregulation of its expression in pulmonary arterial hypertension. Circulation 135(12):1160–1173. https://doi.org/10.1161/CIRCULATIONAHA.116.023218
219. de Jesus Perez V, Yuan K, Alastalo TP, Spiekerkoetter E, Rabinovitch M (2014) Targeting the Wnt signaling pathways in pulmonary arterial hypertension. Drug Discov Today 19 (8):1270–1276. https://doi.org/10.1016/j.drudis.2014.06.014
220. Fan Y, Potdar AA, Gong Y, Eswarappa SM, Donnola S, Lathia JD, Hambardzumyan D, Rich JN, Fox PL (2014) Profilin-1 phosphorylation directs angiocrine expression and glioblastoma progression through HIF-1alpha accumulation. Nat Cell Biol 16(5):445–456. https://doi.org/10.1038/ncb2954
221. Cao Z, Ding BS, Guo P, Lee SB, Butler JM, Casey SC, Simons M, Tam W, Felsher DW, Shido K, Rafii A, Scandura JM, Rafii S (2014) Angiocrine factors deployed by tumor vascular niche induce B cell lymphoma invasiveness and chemoresistance. Cancer Cell 25(3):350–365. https://doi.org/10.1016/j.ccr.2014.02.005
222. Winkler F, Kozin SV, Tong RT, Chae SS, Booth MF, Garkavtsev I, Xu L, Hicklin DJ, Fukumura D, di Tomaso E, Munn LL, Jain RK (2004) Kinetics of vascular normalization

by VEGFR2 blockade governs brain tumor response to radiation: role of oxygenation, angiopoietin-1, and matrix metalloproteinases. Cancer Cell 6(6):553–563. https://doi.org/10.1016/j.ccr.2004.10.011
223. Watkins S, Robel S, Kimbrough IF, Robert SM, Ellis-Davies G, Sontheimer H (2014) Disruption of astrocyte-vascular coupling and the blood-brain barrier by invading glioma cells. Nat Commun 5:4196. https://doi.org/10.1038/ncomms5196
224. Cuddapah VA, Robel S, Watkins S, Sontheimer H (2014) A neurocentric perspective on glioma invasion. Nat Rev Neurosci 15(7):455–465. https://doi.org/10.1038/nrn3765
225. Bikfalvi A, Moenner M, Javerzat S, North S, Hagedorn M (2011) Inhibition of angiogenesis and the angiogenesis/invasion shift. Biochem Soc Trans 39(6):1560–1564. https://doi.org/10.1042/BST20110710
226. Scherer HJ (1938) Structural development in gliomas. Am J Cancer 34:333–351
227. Talasila KM, Soentgerath A, Euskirchen P, Rosland GV, Wang J, Huszthy PC, Prestegarden L, Skaftnesmo KO, Sakariassen PO, Eskilsson E, Stieber D, Keunen O, Brekka N, Moen I, Nigro JM, Vintermyr OK, Lund-Johansen M, Niclou S, Mork SJ, Enger PO, Bjerkvig R, Miletic H (2013) EGFR wild-type amplification and activation promote invasion and development of glioblastoma independent of angiogenesis. Acta Neuropathol 125(5):683–698. https://doi.org/10.1007/s00401-013-1101-1
228. Lu KV, Chang JP, Parachoniak CA, Pandika MM, Aghi MK, Meyronet D, Isachenko N, Fouse SD, Phillips JJ, Cheresh DA, Park M, Bergers G (2012) VEGF inhibits tumor cell invasion and mesenchymal transition through a MET/VEGFR2 complex. Cancer Cell 22(1):21–35. https://doi.org/10.1016/j.ccr.2012.05.037
229. Auf G, Jabouille A, Guerit S, Pineau R, Delugin M, Bouchecareilh M, Magnin N, Favereaux A, Maitre M, Gaiser T, von Deimling A, Czabanka M, Vajkoczy P, Chevet E, Bikfalvi A, Moenner M (2010) Inositol-requiring enzyme 1alpha is a key regulator of angiogenesis and invasion in malignant glioma. Proc Natl Acad Sci U S A 107(35):15553–15558. https://doi.org/10.1073/pnas.0914072107
230. Boye K, Pujol N, D Alves I, Chen YP, Daubon T, Lee YZ, Dedieu S, Constantin M, Bello L, Rossi M, Bjerkvig R, Sue SC, Bikfalvi A, Billottet C (2017) The role of CXCR3/LRP1 cross-talk in the invasion of primary brain tumors. Nat Commun 8(1):1571. https://doi.org/10.1038/s41467-017-01686-y
231. Montana V, Sontheimer H (2011) Bradykinin promotes the chemotactic invasion of primary brain tumors. J Neurosci 31(13):4858–4867. https://doi.org/10.1523/JNEUROSCI.3825-10.2011
232. Moussion C, Girard JP (2011) Dendritic cells control lymphocyte entry to lymph nodes through high endothelial venules. Nature 479(7374):542–546. https://doi.org/10.1038/nature10540
233. Martinet L, Garrido I, Filleron T, Le Guellec S, Bellard E, Fournie JJ, Rochaix P, Girard JP (2011) Human solid tumors contain high endothelial venules: association with T- and B-lymphocyte infiltration and favorable prognosis in breast cancer. Cancer Res 71(17):5678–5687. https://doi.org/10.1158/0008-5472.CAN-11-0431
234. Yao L, Sgadari C, Furuke K, Bloom ET, Teruya-Feldstein J, Tosato G (1999) Contribution of natural killer cells to inhibition of angiogenesis by interleukin-12. Blood 93(5):1612–1621
235. Martinet L, Garrido I, Girard JP (2012) Tumor high endothelial venules (HEVs) predict lymphocyte infiltration and favorable prognosis in breast cancer. Oncoimmunology 1(5):789–790. https://doi.org/10.4161/onci.19787
236. Allen E, Jabouille A, Rivera LB, Lodewijckx I, Missiaen R, Steri V, Feyen K, Tawney J, Hanahan D, Michael IP, Bergers G (2017) Combined antiangiogenic and anti-PD-L1 therapy stimulates tumor immunity through HEV formation. Sci Transl Med 9(385). https://doi.org/10.1126/scitranslmed.aak9679
237. Jain RK (2001) Normalizing tumor vasculature with anti-angiogenic therapy: a new paradigm for combination therapy. Nat Med 7(9):987–989. https://doi.org/10.1038/nm0901-987

238. Jain RK (2003) Molecular regulation of vessel maturation. Nat Med 9(6):685–693. https://doi.org/10.1038/nm0603-685
239. Javerzat SGV, Bikfalvi A (2013) Balancing risks and benefits of anti-angiogenic drugs for malignant glioma. Future Neurol 8(2):159–174
240. Le Noble F, Moyon D, Pardanaud L, Yuan L, Djonov V, Matthijsen R, Breant C, Fleury V, Eichmann A (2004) Flow regulates arterial-venous differentiation in the chick embryo yolk sac. Development 131(2):361–375. https://doi.org/10.1242/dev.00929
241. Hagedorn M, Javerzat S, Gilges D, Meyre A, de Lafarge B, Eichmann A, Bikfalvi A (2005) Accessing key steps of human tumor progression in vivo by using an avian embryo model. Proc Natl Acad Sci U S A 102(5):1643–1648. https://doi.org/10.1073/pnas.0408622102
242. Dumartin L, Quemener C, Laklai H, Herbert J, Bicknell R, Bousquet C, Pyronnet S, Castronovo V, Schilling MK, Bikfalvi A, Hagedorn M (2010) Netrin-1 mediates early events in pancreatic adenocarcinoma progression, acting on tumor and endothelial cells. Gastroenterology 138(4):1595–1606, 1606 e1591–1598. https://doi.org/10.1053/j.gastro.2009.12.061
243. Hagedorn M, Balke M, Schmidt A, Bloch W, Kurz H, Javerzat S, Rousseau B, Wilting J, Bikfalvi A (2004) VEGF coordinates interaction of pericytes and endothelial cells during vasculogenesis and experimental angiogenesis. Dev Dyn 230(1):23–33. https://doi.org/10.1002/dvdy.20020
244. Soulet F, Kilarski WW, Roux-Dalvai F, Herbert JM, Sacewicz I, Mouton-Barbosa E, Bicknell R, Lalor P, Monsarrat B, Bikfalvi A (2013) Mapping the extracellular and membrane proteome associated with the vasculature and the stroma in the embryo. Mol Cell Proteomics 12(8):2293–2312. https://doi.org/10.1074/mcp.M112.024075
245. Saidi A, Javerzat S, Bellahcene A, De Vos J, Bello L, Castronovo V, Deprez M, Loiseau H, Bikfalvi A, Hagedorn M (2008) Experimental anti-angiogenesis causes upregulation of genes associated with poor survival in glioblastoma. Int J Cancer/Journal international du cancer 122 (10):2187–2198. https://doi.org/10.1002/ijc.23313
246. Javerzat S, Franco M, Herbert J, Platonova N, Peille AL, Pantesco V, De Vos J, Assou S, Bicknell R, Bikfalvi A, Hagedorn M (2009) Correlating global gene regulation to angiogenesis in the developing chick extra-embryonic vascular system. PLoS One 4(11):e7856. https://doi.org/10.1371/journal.pone.0007856
247. Cavill R, Sidhu JK, Kilarski W, Javerzat S, Hagedorn M, Ebbels TM, Bikfalvi A, Keun HC (2010) A combined metabonomic and transcriptomic approach to investigate metabolism during development in the chick chorioallantoic membrane. J Proteome Res 9 (6):3126–3134. https://doi.org/10.1021/pr100033t
248. Quemener C, Baud J, Boye K, Dubrac A, Billottet C, Soulet F, Darlot F, Dumartin L, Sire M, Grepin R, Daubon T, Rayne F, Wodrich H, Couvelard A, Pineau R, Schilling M, Castronovo V, Sue SC, Clarke K, Lomri A, Khatib AM, Hagedorn M, Prats H, Bikfalvi A (2016) Dual roles for CXCL4 chemokines and CXCR3 in angiogenesis and invasion of pancreatic cancer. Cancer Res 76(22):6507–6519. https://doi.org/10.1158/0008-5472.CAN-15-2864
249. Russell WMS, Burch RL (1959) The principles of humane experimental technique. Methuen, London
250. Folkman J, Watson K, Ingber D, Hanahan D (1989) Induction of angiogenesis during the transition from hyperplasia to neoplasia. Nature 339(6219):58–61. https://doi.org/10.1038/339058a0
251. Ferrara N, Kerbel RS (2005) Angiogenesis as a therapeutic target. Nature 438(7070):967–974. https://doi.org/10.1038/nature04483
252. Ebos JM, Kerbel RS (2011) Antiangiogenic therapy: impact on invasion, disease progression, and metastasis. Nat Rev Clin Oncol 8(4):210–221. https://doi.org/10.1038/nrclinonc.2011.21
253. Carmeliet P, Jain RK (2011) Principles and mechanisms of vessel normalization for cancer and other angiogenic diseases. Nat Rev Drug Discov 10(6):417–427. https://doi.org/10.1038/nrd3455
254. Shaked Y, Henke E, Roodhart JM, Mancuso P, Langenberg MH, Colleoni M, Daenen LG, Man S, Xu P, Emmenegger U, Tang T, Zhu Z, Witte L, Strieter RM, Bertolini F, Voest EE, Benezra R, Kerbel RS (2008) Rapid chemotherapy-induced acute endothelial progenitor cell

mobilization: implications for antiangiogenic drugs as chemosensitizing agents. Cancer Cell 14(3):263–273. https://doi.org/10.1016/j.ccr.2008.08.001
255. Jain RK (2005) Normalization of tumor vasculature: an emerging concept in antiangiogenic therapy. Science 307(5706):58–62. https://doi.org/10.1126/science.1104819
256. Keunen O, Johansson M, Oudin A, Sanzey M, Rahim SA, Fack F, Thorsen F, Taxt T, Bartos M, Jirik R, Miletic H, Wang J, Stieber D, Stuhr L, Moen I, Rygh CB, Bjerkvig R, Niclou SP (2011) Anti-VEGF treatment reduces blood supply and increases tumor cell invasion in glioblastoma. Proc Natl Acad Sci U S A 108(9):3749–3754. https://doi.org/10.1073/pnas.1014480108
257. Fack F, Espedal H, Keunen O, Golebiewska A, Obad N, Harter PN, Mittelbronn M, Bahr O, Weyerbrock A, Stuhr L, Miletic H, Sakariassen PO, Stieber D, Rygh CB, Lund-Johansen M, Zheng L, Gottlieb E, Niclou SP, Bjerkvig R (2015) Bevacizumab treatment induces metabolic adaptation toward anaerobic metabolism in glioblastomas. Acta Neuropathol 129(1):115–131. https://doi.org/10.1007/s00401-014-1352-5
258. LeBleu VS, O'Connell JT, Gonzalez Herrera KN, Wikman H, Pantel K, Haigis MC, de Carvalho FM, Damascena A, Domingos Chinen LT, Rocha RM, Asara JM, Kalluri R (2014) PGC-1alpha mediates mitochondrial biogenesis and oxidative phosphorylation in cancer cells to promote metastasis. Nat Cell Biol 16(10):992–1003, 1001–1015. https://doi.org/10.1038/ncb3039
259. Ferrara N (2010) Vascular endothelial growth factor and age-related macular degeneration: from basic science to therapy. Nat Med 16(10):1107–1111. https://doi.org/10.1038/nm1010-1107
260. Jiang S, Park C, Barner JC (2014) Ranibizumab for age-related macular degeneration: a meta-analysis of dose effects and comparison with no anti-VEGF treatment and bevacizumab. J Clin Pharm Ther 39(3):234–239. https://doi.org/10.1111/jcpt.12146
261. Yang J, Wang X, Fuh G, Yu L, Wakshull E, Khosraviani M, Day ES, Demeule B, Liu J, Shire SJ, Ferrara N, Yadav S (2014) Comparison of binding characteristics and in vitro activities of three inhibitors of vascular endothelial growth factor A. Mol Pharm 11(10):3421–3430. https://doi.org/10.1021/mp500160v
262. Teesalu T, Sugahara KN, Ruoslahti E (2012) Mapping of vascular ZIP codes by phage display. Methods Enzymol 503:35–56. https://doi.org/10.1016/B978-0-12-396962-0.00002-1
263. Staquicini FI, Moeller BJ, Arap W, Pasqualini R (2010) Combinatorial vascular targeting in translational medicine. Proteomics Clin Appl 4(6-7):626–632. https://doi.org/10.1002/prca.200900213
264. Elia G, Fugmann T, Neri D (2014) From target discovery to clinical trials with armed antibody products. J Proteome 107:50–55. https://doi.org/10.1016/j.jprot.2014.02.034
265. Wigle JT, Oliver G (1999) Prox1 function is required for the development of the murine lymphatic system. Cell 98(6):769–778
266. Nicenboim J, Malkinson G, Lupo T, Asaf L, Sela Y, Mayseless O, Gibbs-Bar L, Senderovich N, Hashimshony T, Shin M, Jerafi-Vider A, Avraham-Davidi I, Krupalnik V, Hofi R, Almog G, Astin JW, Golani O, Ben-Dor S, Crosier PS, Herzog W, Lawson ND, Hanna JH, Yanai I, Yaniv K (2015) Lymphatic vessels arise from specialized angioblasts within a venous niche. Nature 522(7554):56–61. https://doi.org/10.1038/nature14425
267. Martinez-Corral I, Ulvmar MH, Stanczuk L, Tatin F, Kizhatil K, John SW, Alitalo K, Ortega S, Makinen T (2015) Nonvenous origin of dermal lymphatic vasculature. Circ Res 116(10):1649–1654. https://doi.org/10.1161/CIRCRESAHA.116.306170
268. Klotz L, Norman S, Vieira JM, Masters M, Rohling M, Dube KN, Bollini S, Matsuzaki F, Carr CA, Riley PR (2015) Cardiac lymphatics are heterogeneous in origin and respond to injury. Nature 522(7554):62–67. https://doi.org/10.1038/nature14483
269. Skobe M, Hawighorst T, Jackson DG, Prevo R, Janes L, Velasco P, Riccardi L, Alitalo K, Claffey K, Detmar M (2001) Induction of tumor lymphangiogenesis by VEGF-C promotes breast cancer metastasis. Nat Med 7(2):192–198. https://doi.org/10.1038/84643

270. Karpanen T, Egeblad M, Karkkainen MJ, Kubo H, Yla-Herttuala S, Jaattela M, Alitalo K (2001) Vascular endothelial growth factor C promotes tumor lymphangiogenesis and intralymphatic tumor growth. Cancer Res 61(5):1786–1790
271. Vaahtomeri K, Karaman S, Makinen T, Alitalo K (2017) Lymphangiogenesis guidance by paracrine and pericellular factors. Genes Dev 31(16):1615–1634. https://doi.org/10.1101/gad.303776.117
272. Deng Y, Atri D, Eichmann A, Simons M (2013) Endothelial ERK signaling controls lymphatic fate specification. J Clin Invest 123(3):1202–1215. https://doi.org/10.1172/JCI63034
273. Cao R, Ji H, Feng N, Zhang Y, Yang X, Andersson P, Sun Y, Tritsaris K, Hansen AJ, Dissing S, Cao Y (2012) Collaborative interplay between FGF-2 and VEGF-C promotes lymphangiogenesis and metastasis. Proc Natl Acad Sci U S A 109(39):15894–15899. https://doi.org/10.1073/pnas.1208324109
274. Sabine A, Bovay E, Demir CS, Kimura W, Jaquet M, Agalarov Y, Zangger N, Scallan JP, Graber W, Gulpinar E, Kwak BR, Makinen T, Martinez-Corral I, Ortega S, Delorenzi M, Kiefer F, Davis MJ, Djonov V, Miura N, Petrova TV (2015) FOXC2 and fluid shear stress stabilize postnatal lymphatic vasculature. J Clin Invest 125(10):3861–3877. https://doi.org/10.1172/JCI80454
275. Gordon K, Schulte D, Brice G, Simpson MA, Roukens MG, van Impel A, Connell F, Kalidas K, Jeffery S, Mortimer PS, Mansour S, Schulte-Merker S, Ostergaard P (2013) Mutation in vascular endothelial growth factor-C, a ligand for vascular endothelial growth factor receptor-3, is associated with autosomal dominant milroy-like primary lymphedema. Circ Res 112(6):956–960. https://doi.org/10.1161/CIRCRESAHA.113.300350
276. Karaman S, Detmar M (2014) Mechanisms of lymphatic metastasis. J Clin Invest 124 (3):922–928. https://doi.org/10.1172/JCI71606
277. Dadras SS, Lange-Asschenfeldt B, Velasco P, Nguyen L, Vora A, Muzikansky A, Jahnke K, Hauschild A, Hirakawa S, Mihm MC, Detmar M (2005) Tumor lymphangiogenesis predicts melanoma metastasis to sentinel lymph nodes. Mod Pathol 18(9):1232–1242. https://doi.org/10.1038/modpathol.3800410
278. Mandriota SJ, Jussila L, Jeltsch M, Compagni A, Baetens D, Prevo R, Banerji S, Huarte J, Montesano R, Jackson DG, Orci L, Alitalo K, Christofori G, Pepper MS (2001) Vascular endothelial growth factor-C-mediated lymphangiogenesis promotes tumour metastasis. EMBO J 20(4):672–682. https://doi.org/10.1093/emboj/20.4.672
279. Takahashi T, Kalka C, Masuda H, Chen D, Silver M, Kearney M, Magner M, Isner JM, Asahara T (1999) Ischemia- and cytokine-induced mobilization of bone marrow-derived endothelial progenitor cells for neovascularization. Nat Med 5(4):434–438. https://doi.org/10.1038/7434
280. Seandel M, Butler J, Lyden D, Rafii S (2008) A catalytic role for proangiogenic marrow-derived cells in tumor neovascularization. Cancer Cell 13(3):181–183. https://doi.org/10.1016/j.ccr.2008.02.016
281. Bompais H, Chagraoui J, Canron X, Crisan M, Liu XH, Anjo A, Tolla-Le Port C, Leboeuf M, Charbord P, Bikfalvi A, Uzan G (2004) Human endothelial cells derived from circulating progenitors display specific functional properties compared with mature vessel wall endothelial cells. Blood 103(7):2577–2584. https://doi.org/10.1182/blood-2003-08-2770
282. Leroyer AS, Anfosso F, Lacroix R, Sabatier F, Simoncini S, Njock SM, Jourde N, Brunet P, Camoin-Jau L, Sampol J, Dignat-George F (2010) Endothelial-derived microparticles: biological conveyors at the crossroad of inflammation, thrombosis and angiogenesis. Thromb Haemost 104(3):456–463. https://doi.org/10.1160/TH10-02-0111
283. Gao D, Mittal V (2009) The role of bone-marrow-derived cells in tumor growth, metastasis initiation and progression. Trends Mol Med 15(8):333–343. https://doi.org/10.1016/j.molmed.2009.06.006
284. Purhonen S, Palm J, Rossi D, Kaskenpaa N, Rajantie I, Yla-Herttuala S, Alitalo K, Weissman IL, Salven P (2008) Bone marrow-derived circulating endothelial precursors do not contribute

to vascular endothelium and are not needed for tumor growth. Proc Natl Acad Sci U S A 105 (18):6620–6625. https://doi.org/10.1073/pnas.0710516105
285. Kerbel RS, Benezra R, Lyden DC, Hattori K, Heissig B, Nolan DJ, Mittal V, Shaked Y, Dias S, Bertolini F, Rafii S (2008) Endothelial progenitor cells are cellular hubs essential for neoangiogenesis of certain aggressive adenocarcinomas and metastatic transition but not adenomas. Proc Natl Acad Sci U S A 105(34):E54; author reply E55. doi:https://doi.org/10.1073/pnas.0804876105
286. Peters BA, Diaz LA, Polyak K, Meszler L, Romans K, Guinan EC, Antin JH, Myerson D, Hamilton SR, Vogelstein B, Kinzler KW, Lengauer C (2005) Contribution of bone marrow-derived endothelial cells to human tumor vasculature. Nat Med 11(3):261–262. https://doi.org/10.1038/nm1200
287. Ricci-Vitiani L, Pallini R, Biffoni M, Todaro M, Invernici G, Cenci T, Maira G, Parati EA, Stassi G, Larocca LM, De Maria R (2010) Tumour vascularization via endothelial differentiation of glioblastoma stem-like cells. Nature 468(7325):824–828. https://doi.org/10.1038/nature09557
288. Wang R, Chadalavada K, Wilshire J, Kowalik U, Hovinga KE, Geber A, Fligelman B, Leversha M, Brennan C, Tabar V (2010) Glioblastoma stem-like cells give rise to tumour endothelium. Nature 468(7325):829–833. https://doi.org/10.1038/nature09624
289. Golebiewska A, Bougnaud S, Stieber D, Brons NH, Vallar L, Hertel F, Klink B, Schrock E, Bjerkvig R, Niclou SP (2013) Side population in human glioblastoma is non-tumorigenic and characterizes brain endothelial cells. Brain 136(5):1462–1475. https://doi.org/10.1093/brain/awt025
290. Cheng L, Huang Z, Zhou W, Wu Q, Donnola S, Liu JK, Fang X, Sloan AE, Mao Y, Lathia JD, Min W, McLendon RE, Rich JN, Bao S (2013) Glioblastoma stem cells generate vascular pericytes to support vessel function and tumor growth. Cell 153(1):139–152. https://doi.org/10.1016/j.cell.2013.02.021
291. Williamson SC, Metcalf RL, Trapani F, Mohan S, Antonello J, Abbott B, Leong HS, Chester CP, Simms N, Polanski R, Nonaka D, Priest L, Fusi A, Carlsson F, Carlsson A, Hendrix MJ, Seftor RE, Seftor EA, Rothwell DG, Hughes A, Hicks J, Miller C, Kuhn P, Brady G, Simpson KL, Blackhall FH, Dive C (2016) Vasculogenic mimicry in small cell lung cancer. Nat Commun 7:13322. https://doi.org/10.1038/ncomms13322
292. Sanchez-Duffhues G, Orlova V, Ten Dijke P (2016) In Brief: Endothelial-to-mesenchymal transition. J Pathol 238(3):378–380. https://doi.org/10.1002/path.4653
293. Arderiu G, Espinosa S, Pena E, Crespo J, Aledo R, Bogdanov VY, Badimon L (2017) Tissue factor variants induce monocyte transformation and transdifferentiation into endothelial cell-like cells. J Thromb Haemost 15(8):1689–1703. https://doi.org/10.1111/jth.13751
294. Shiler C (1901) The nerves of the capillaries, with remarks on nerve-endings in muscle. A new theory of lymph-formation and glandular secretion. J Exp Med:493–512
295. Honma Y, Araki T, Gianino S, Bruce A, Heuckeroth R, Johnson E, Milbrandt J (2002) Artemin is a vascular-derived neurotropic factor for developing sympathetic neurons. Neuron 35(2):267–282
296. Soker S, Takashima S, Miao HQ, Neufeld G, Klagsbrun M (1998) Neuropilin-1 is expressed by endothelial and tumor cells as an isoform-specific receptor for vascular endothelial growth factor. Cell 92(6):735–745
297. Jones EA, Yuan L, Breant C, Watts RJ, Eichmann A (2008) Separating genetic and hemodynamic defects in neuropilin 1 knockout embryos. Development 135(14):2479–2488. https://doi.org/10.1242/dev.014902
298. Bouvree K, Larrivee B, Lv X, Yuan L, DeLafarge B, Freitas C, Mathivet T, Breant C, Tessier-Lavigne M, Bikfalvi A, Eichmann A, Pardanaud L (2008) Netrin-1 inhibits sprouting angiogenesis in developing avian embryos. Dev Biol 318(1):172–183. https://doi.org/10.1016/j.ydbio.2008.03.023
299. Larrivee B, Freitas C, Trombe M, Lv X, Delafarge B, Yuan L, Bouvree K, Breant C, Del Toro R, Brechot N, Germain S, Bono F, Dol F, Claes F, Fischer C, Autiero M, Thomas JL,

Carmeliet P, Tessier-Lavigne M, Eichmann A (2007) Activation of the UNC5B receptor by Netrin-1 inhibits sprouting angiogenesis. Genes Dev 21(19):2433–2447. https://doi.org/10.1101/gad.437807

300. Koch AW, Mathivet T, Larrivee B, Tong RK, Kowalski J, Pibouin-Fragner L, Bouvree K, Stawicki S, Nicholes K, Rathore N, Scales SJ, Luis E, del Toro R, Freitas C, Breant C, Michaud A, Corvol P, Thomas JL, Wu Y, Peale F, Watts RJ, Tessier-Lavigne M, Bagri A, Eichmann A (2011) Robo4 maintains vessel integrity and inhibits angiogenesis by interacting with UNC5B. Dev Cell 20(1):33–46. https://doi.org/10.1016/j.devcel.2010.12.001
301. Adams RH, Eichmann A (2010) Axon guidance molecules in vascular patterning. Cold Spring Harb Perspect Biol 2(5):a001875. https://doi.org/10.1101/cshperspect.a001875
302. Lange C, Storkebaum E, de Almodovar CR, Dewerchin M, Carmeliet P (2016) Vascular endothelial growth factor: a neurovascular target in neurological diseases. Nat Rev Neurol 12 (8):439–454. https://doi.org/10.1038/nrneurol.2016.88
303. Guerit S, Allain AE, Leon C, Cazenave W, Ferrara N, Branchereau P, Bikfalvi A (2014) VEGF modulates synaptic activity in the developing spinal cord. Dev Neurobiol 74 (11):1110–1122. https://doi.org/10.1002/dneu.22187
304. Cao L, Jiao X, Zuzga DS, Liu Y, Fong DM, Young D, During MJ (2004) VEGF links hippocampal activity with neurogenesis, learning and memory. Nat Genet 36(8):827–835. https://doi.org/10.1038/ng1395
305. Brunet I, Gordon E, Han J, Cristofaro B, Broqueres-You D, Liu C, Bouvree K, Zhang J, del Toro R, Mathivet T, Larrivee B, Jagu J, Pibouin-Fragner L, Pardanaud L, Machado MJ, Kennedy TE, Zhuang Z, Simons M, Levy BI, Tessier-Lavigne M, Grenz A, Eltzschig H, Eichmann A (2014) Netrin-1 controls sympathetic arterial innervation. J Clin Invest 124 (7):3230–3240. https://doi.org/10.1172/JCI75181
306. Nam J, Onitsuka I, Hatch J, Uchida Y, Ray S, Huang S, Li W, Zang H, Ruiz-Lozano P, Mukouyama YS (2013) Coronary veins determine the pattern of sympathetic innervation in the developing heart. Development 140(7):1475–1485. https://doi.org/10.1242/dev.087601
307. Renault MA, Chapouly C, Yao Q, Larrieu-Lahargue F, Vandierdonck S, Reynaud A, Petit M, Jaspard-Vinassa B, Belloc I, Traiffort E, Ruat M, Duplaa C, Couffinhal T, Desgranges C, Gadeau AP (2013) Desert hedgehog promotes ischemia-induced angiogenesis by ensuring peripheral nerve survival. Circ Res 112(5):762–770. https://doi.org/10.1161/CIRCRESAHA.113.300871
308. Park BH, Song KJ, Yoon SJ, Park HS, Jang KY, Zhou L, Lee SY, Lee KB, Kim JR (2011) Acceleration of spinal fusion using COMP-angiopoietin 1 with allografting in a rat model. Bone 49(3):447–454. https://doi.org/10.1016/j.bone.2011.05.020
309. Kosacka J, Nowicki M, Kloting N, Kern M, Stumvoll M, Bechmann I, Serke H, Bluher M (2012) COMP-angiopoietin-1 recovers molecular biomarkers of neuropathy and improves vascularisation in sciatic nerve of ob/ob mice. PLoS One 7(3):e32881. https://doi.org/10.1371/journal.pone.0032881
310. He L, Vanlandewijck M, Raschperger E, Andaloussi Mae M, Jung B, Lebouvier T, Ando K, Hofmann J, Keller A, Betsholtz C (2016) Analysis of the brain mural cell transcriptome. Sci Rep 6:35108. https://doi.org/10.1038/srep35108
311. Roviello G, Bachelot T, Hudis CA, Curigliano G, Reynolds AR, Petrioli R, Generali D (2017) The role of bevacizumab in solid tumours: a literature based meta-analysis of randomised trials. Eur J Cancer 75:245–258. https://doi.org/10.1016/j.ejca.2017.01.026
312. Ranpura V, Hapani S, Wu S (2011) Treatment-related mortality with bevacizumab in cancer patients: a meta-analysis. JAMA 305(5):487–494. https://doi.org/10.1001/jama.2011.51
313. Arrillaga-Romany I, Reardon DA, Wen PY (2014) Current status of antiangiogenic therapies for glioblastomas. Expert Opin Investig Drugs 23(2):199–210. https://doi.org/10.1517/13543784.2014.856880
314. Fathpour P, Obad N, Espedal H, Stieber D, Keunen O, Sakariassen PO, Niclou SP, Bjerkvig R (2014) Bevacizumab treatment for human glioblastoma. Can it induce cognitive impairment? Neuro-Oncology 16(5):754–756. https://doi.org/10.1093/neuonc/nou013
315. Paez-Ribes M, Allen E, Hudock J, Takeda T, Okuyama H, Vinals F, Inoue M, Bergers G, Hanahan D, Casanovas O (2009) Antiangiogenic therapy elicits malignant progression of

tumors to increased local invasion and distant metastasis. Cancer Cell 15(3):220–231. https://doi.org/10.1016/j.ccr.2009.01.027

316. Clasper S, Royston D, Baban D, Cao Y, Ewers S, Butz S, Vestweber D, Jackson DG (2008) A novel gene expression profile in lymphatics associated with tumor growth and nodal metastasis. Cancer Res 68(18):7293–7303. https://doi.org/10.1158/0008-5472.CAN-07-6506
317. Dieterich LC, Seidel CD, Detmar M (2014) Lymphatic vessels: new targets for the treatment of inflammatory diseases. Angiogenesis 17(2):359–371. https://doi.org/10.1007/s10456-013-9406-1
318. Henri O, Pouehe C, Houssari M, Galas L, Nicol L, Edwards-Levy F, Henry JP, Dumesnil A, Boukhalfa I, Banquet S, Schapman D, Thuillez C, Richard V, Mulder P, Brakenhielm E (2016) Selective stimulation of cardiac lymphangiogenesis reduces myocardial edema and fibrosis leading to improved cardiac function following myocardial infarction. Circulation 133 (15):1484–1497.; discussion 1497. https://doi.org/10.1161/CIRCULATIONAHA.115.020143
319. Jain RK, Duda DG, Willett CG, Sahani DV, Zhu AX, Loeffler JS, Batchelor TT, Sorensen AG (2009) Biomarkers of response and resistance to antiangiogenic therapy. Nat Rev Clin Oncol 6 (6):327–338. https://doi.org/10.1038/nrclinonc.2009.63
320. Sessa C, Guibal A, Del Conte G, Ruegg C (2008) Biomarkers of angiogenesis for the development of antiangiogenic therapies in oncology: tools or decorations? Nat Clin Pract Oncol 5(7):378–391. https://doi.org/10.1038/ncponc1150
321. Jubb AM, Harris AL (2010) Biomarkers to predict the clinical efficacy of bevacizumab in cancer. Lancet Oncol 11(12):1172–1183. https://doi.org/10.1016/S1470-2045(10)70232-1
322. Pohl M, Werner N, Munding J, Tannapfel A, Graeven U, Nickenig G, Schmiegel W, Reinacher-Schick A (2011) Biomarkers of anti-angiogenic therapy in metastatic colorectal cancer (mCRC): original data and review of the literature. Zeitschrift fur Gastroenterologie 49 (10):1398–1406. https://doi.org/10.1055/s-0031-1281752
323. Schneider BP, Radovich M, Sledge GW, Robarge JD, Li L, Storniolo AM, Lemler S, Nguyen AT, Hancock BA, Stout M, Skaar T, Flockhart DA (2008) Association of polymorphisms of angiogenesis genes with breast cancer. Breast Cancer Res Treat 111(1):157-163. doi:https://doi.org/10.1007/s10549-007-9755-9
324. Willett CG, Duda DG, di Tomaso E, Boucher Y, Ancukiewicz M, Sahani DV, Lahdenranta J, Chung DC, Fischman AJ, Lauwers GY, Shellito P, Czito BG, Wong TZ, Paulson E, Poleski M, Vujaskovic Z, Bentley R, Chen HX, Clark JW, Jain RK (2009) Efficacy, safety, and biomarkers of neoadjuvant bevacizumab, radiation therapy, and fluorouracil in rectal cancer: a multidisciplinary phase II study. J Clin Oncol 27(18):3020–3026. https://doi.org/10.1200/JCO.2008.21.1771
325. Emblem KE, Farrar CT, Gerstner ER, Batchelor TT, Borra RJ, Rosen BR, Sorensen AG, Jain RK (2014) Vessel caliber–a potential MRI biomarker of tumour response in clinical trials. Nat Rev Clin Oncol 11(10):566–584. https://doi.org/10.1038/nrclinonc.2014.126
326. Daniels LB, Maisel AS (2015) Cardiovascular biomarkers and sex: the case for women. Nat Rev Cardiol. https://doi.org/10.1038/nrcardio.2015.105
327. Awata T (2010) Vascular endothelial growth factor gene polymorphisms in susceptibility to coronary artery disease. Am J Hypertens 23(9):938–939. https://doi.org/10.1038/ajh.2010.151
328. Bry M, Kivela R, Leppanen VM, Alitalo K (2014) Vascular endothelial growth factor-B in physiology and disease. Physiol Rev 94(3):779–794. https://doi.org/10.1152/physrev.00028.2013
329. De Sutter J, Van de Veire NR, Struyf S, Philippe J, De Buyzere M, Van Damme J (2012) PF-4var/CXCL4L1 predicts outcome in stable coronary artery disease patients with preserved left ventricular function. PLoS One 7(2):e31343. https://doi.org/10.1371/journal.pone.0031343
330. Joshi S, Viljoen A (2015) Renal biomarkers for the prediction of cardiovascular disease. Curr Opin Cardiol 30(4):454–460. https://doi.org/10.1097/HCO.0000000000000177

331. Shah SH, Newgard CB (2015) Integrated metabolomics and genomics: systems approaches to biomarkers and mechanisms of cardiovascular disease. Circ Cardiovasc Genet 8(2):410–419. https://doi.org/10.1161/CIRCGENETICS.114.000223
332. Saif J, Emanueli C (2014) miRNAs in post-ischaemic angiogenesis and vascular remodelling. Biochem Soc Trans 42(6):1629–1636. https://doi.org/10.1042/BST20140263
333. Pecot CV, Rupaimoole R, Yang D, Akbani R, Ivan C, Lu C, Wu S, Han HD, Shah MY, Rodriguez-Aguayo C, Bottsford-Miller J, Liu Y, Kim SB, Unruh A, Gonzalez-Villasana V, Huang L, Zand B, Moreno-Smith M, Mangala LS, Taylor M, Dalton HJ, Sehgal V, Wen Y, Kang Y, Baggerly KA, Lee JS, Ram PT, Ravoori MK, Kundra V, Zhang X, Ali-Fehmi R, Gonzalez-Angulo AM, Massion PP, Calin GA, Lopez-Berestein G, Zhang W, Sood AK (2013) Tumour angiogenesis regulation by the miR-200 family. Nat Commun 4:2427. https://doi.org/10.1038/ncomms3427
334. Joosten SC, Hamming L, Soetekouw PM, Aarts MJ, Veeck J, van Engeland M, Tjan-Heijnen VC (2015) Resistance to sunitinib in renal cell carcinoma: From molecular mechanisms to predictive markers and future perspectives. Biochim Biophys Acta 1855(1):1–16. https://doi.org/10.1016/j.bbcan.2014.11.002
335. Romaine SP, Tomaszewski M, Condorelli G, Samani NJ (2015) MicroRNAs in cardiovascular disease: an introduction for clinicians. Heart 101(12):921–928. https://doi.org/10.1136/heartjnl-2013-305402
336. Michalik KM, You X, Manavski Y, Doddaballapur A, Zornig M, Braun T, John D, Ponomareva Y, Chen W, Uchida S, Boon RA, Dimmeler S (2014) Long noncoding RNA MALAT1 regulates endothelial cell function and vessel growth. Circ Res 114(9):1389–1397. https://doi.org/10.1161/CIRCRESAHA.114.303265
337. Yan B, Yao J, Liu JY, Li XM, Wang XQ, Li YJ, Tao ZF, Song YC, Chen Q, Jiang Q (2015) lncRNA-MIAT regulates microvascular dysfunction by functioning as a competing endogenous RNA. Circ Res 116(7):1143–1156. https://doi.org/10.1161/CIRCRESAHA.116.305510
338. Kuhn TS (1996) The structure of scientific revolutions, 3rd edn. University of Chicago Press, Chicago, IL
339. Kuhn TS (1962) Historical structure of scientific discovery. Science 136(3518):760–764
340. Ayres P (2012) The life of Arthur Tansley. Wiley-Blackwell, Oxford
341. Hinman VF, Davidson EH (2007) Evolutionary plasticity of developmental gene regulatory network architecture. Proc Natl Acad Sci U S A 104(49):19404–19409. https://doi.org/10.1073/pnas.0709994104
342. Munoz-Chapuli R (2011) Evolution of angiogenesis. Int J Dev Biol 55(4-5):345–351. https://doi.org/10.1387/ijdb.103212rm
343. Popper K (1935) Logik der Forschung. Julius Springer, Wien, Östereich
344. Benveniste J (1988) Benveniste on the Benveniste affair. Nature 335(6193):759. https://doi.org/10.1038/335759a0
345. Montagnier L, Del Giudice E, Aissa J, Lavallee C, Motschwiller S, Capolupo A, Polcari A, Romano P, Tedeschi A, Vitiello G (2015) Transduction of DNA information through water and electromagnetic waves. Electromag Biol Med 34(2):106–112. https://doi.org/10.3109/15368378.2015.1036072
346. Seralini GE, Clair E, Mesnage R, Gress S, Defarge N, Malatesta M, Hennequin D, de Vendomois JS (2012) Long term toxicity of a Roundup herbicide and a Roundup-tolerant genetically modified maize. Food Chem Toxicol 50(11):4221–4231. https://doi.org/10.1016/j.fct.2012.08.005
347. Pradeu T, Jaeger S, Vivier E (2013) The speed of change: towards a discontinuity theory of immunity? Nat Rev Immunol 13(10):764–769. https://doi.org/10.1038/nri3521
348. Jain H, Jackson T (2017) Mathematical modeling of cellular cross-talk between endothelial and tumor cells highlights counterintuitive effects of VEGF-targeted therapies. Bull Math Biol. https://doi.org/10.1007/s11538-017-0273-6
349. Grogan JA, Connor AJ, Markelc B, Muschel RJ, Maini PK, Byrne HM, Pitt-Francis JM (2017) Microvessel chaste: an open library for spatial modeling of vascularized tissues. Biophys J 112(9):1767–1772. https://doi.org/10.1016/j.bpj.2017.03.036

Index

A. Bikfalvi, *A Brief History of Blood and Lymphatic Vessels*,
https://doi.org/10.1007/978-3-319-74376-9

MIX
Papier aus verantwortungsvollen Quellen
Paper from responsible sources
FSC® C105338
FSC
www.fsc.org

If you have any concerns about our products,
you can contact us on
ProductSafety@springernature.com

In case Publisher is established outside the EU,
the EU authorized representative is:
Springer Nature Customer Service Center GmbH
Europaplatz 3, 69115 Heidelberg, Germany

Printed by Libri Plureos GmbH
in Hamburg, Germany